I0019521

JOURNAL OF CYBER SECURITY AND MOBILITY

Volume 3, No. 1 (January 2014)

Special Issue on
*Intelligent Data Acquisition and
Advanced Computing Systems: Technology
and Applications (IDAACS'2013)*

Guest Editors:
Igor Kotenko and George Markowsky

JOURNAL OF CYBER SECURITY AND MOBILITY

Aim
Journal of Cyber Security and Mobility provides an in-depth and holistic view of security and solutions from practical to theoretical aspects. It covers topics that are equally valuable for practitioners as well as those new in the field.

Scope
The journal covers security issues in cyber space and solutions thereof. As cyber space has moved towards the wireless/mobile world, issues in wireless/mobile communications will also be published. The publication will take a holistic view. Some example topics are: security in mobile networks, security and mobility optimization, cyber security, cloud security, Internet of Things (IoT) and machine-to-machine technologies.

Published, sold and distributed by:
River Publishers
P.O. Box 1657
Algade 42
9000 Aalborg
Denmark

Tel.: +45369953197
www.riverpublishers.com

Journal of Cyber Security and Mobility is published four times a year.
Publication programme, 2014: Volume 3 (1 issues)

ISSN 2245-1439 (Print Version)
ISSN 2245-4578 (Online Version)
ISBN 978-87-93102-96-5 (this issue)

JOURNAL OF CYBER SECURITY AND MOBILITY COMMUNICATIONS

Volume 3, No. 1 (January 2014)

Editorial Foreword: Introduction to the Special Issue of Cyber Security and Mobility on Cyber Security and Software

Cyber security is a hotbed of activity and has become more vital than ever. As the stakes continue to rise more effort is being focused on cyber defense. Of special interest is the creation of better tools for the development of secure software as well as tools that help users find vulnerabilities and better understand the types of attacks that they will have to face.

All the papers in this issue are derived from the best papers presented in the Special Stream at the 7th IEEE International Conference on Intelligent Data Acquisition and Advanced Computing Systems: Technology and Applications (IDAACS'2013) which was held in Berlin, Germany, September 12-14, 2013.

The IDAACS conferences are unique in that their goal is to bring together Western researchers with researchers from the former Soviet Union. As it is well known, there are many strong scholars in the former Soviet Union and it is valuable to have some exposure to their research.

The papers in this issue are:

- *"The Sad History of Random Bits"* by George Markowsky applies the concept of a system accident to the analysis of the problems that have cropped up with pseudo-random number generators over the last 60 years.
- *"Fast Network Attack Modeling and Security Evaluation based on Attack Graphs"* by Igor Kotenko and Andrey Chechulin describes an Attack Modeling and Security Evaluation Component that can construct attack graphs that can be used for network modeling and security evaluations.
- *"Code Search API, base of Parallel Code Refactoring System For Safety Standards Compliance"* by Peter Jurnéčka, Peter Hanáček and Matej Kačic considers a scheme for using verified code patterns in the construction of new software. It discusses the use of this API in aviation and medical software.
- *"Memory Acquisition by Using a Network Card"* by Stefan Balogh of Slovakia outlines a new approach to rootkit detection.
- *"Making Static Code Analysis More Efficient"* by Oksana V. Pomorova and Dmytro O. Ivanchyshyn describes how to use libraries of software vulnerabilities to make static analysis of programs more efficient.

With papers originating from the United States, the Russian Federation, the Czech Republic, Slovakia, and Ukraine this issue provides insight into the research going on across the globe. We hope that you find this special issue of value.

Igor Kotenko
George Markowsky
Co-editors

The Sad History of Random Bits

George Markowsky

School of Computing & Information Science, University of Maine,
markov@maine.edu

Received 5 February 2014; Accepted 27 April 2014;
Publication 2 June 2014

Abstract

In this paper we examine the history of using random numbers in computer programs. Unfortunately, this history is sad because it is replete with disasters ranging from one of the first pseudo-random number generators, RANDU, being very bad to the most recent efforts by the NSA to undermine the pseudo-random number generator in RSA's BSAFE cryptographic library. Failures in this area have been both intentional and unintentional, but unfortunately the same sorts of mistakes are repeated. The repeated failures in getting our "random numbers" correct suggests that there might be some systemic reasons for these failures. In this paper we review some of these failures in more detail, and the 2006 Debian OpenSSL Debacle in great detail. This last event left users of Debian and its derivatives with seriously compromised cryptographic capabilities for two years. We also illustrate how this failure can be exploited in an attack. We also modify the concept of a system accident developed in the work of Charles Perrow [1]. We identify some system failures in building pseudo-random number generators and offer some suggestions to help develop PRNGs and other code more securely.

Keywords: Debian; system accident; SSL; SSH, Bitcoin, cryptography; security breach; software engineering, PRNG, pseudo-random numbers, booby trap, BSAFE, Dual_EC_DRNG.

Journal of Cyber Security, Vol. 3 No. 1 , 1–26.
doi: 10.13052/jcsm2245-1439.311

1 Introduction

In most cases, when something goes wrong, we look for someone to blame. The underlying assumption is that someone must have been the direct cause of each particular mishap. In [1], Charles Perrow introduced the concept of a *system accident*, something he also called a *normal accident*. The basic idea is that while a serious accident might include a number of smaller events the serious accident occurred because the system allowed the smaller events to interact in a manner that combined their contributions. While his analysis targets physical systems, his emphasis on looking at the various components of a system is valuable for virtual systems as well. In this paper, we develop a modified version of a system accident that we think is easier to apply to software systems. We will apply it to a number of incidents involving pseudo-random numbers generators and examine it in detail in the analysis of a little known incident that occurred in 2006 in which the Debian SSL library was compromised. It took two years before someone noticed that the library had been compromised during which time SSL and other services that use the SSL libraries, like SSH, were vulnerable. We conclude with some suggestions on improving the system. This article is a much expanded version of [2].

2 Randomness

People in general are not comfortable with the concept of randomness. People like to feel that things happen for a reason. Sentiments such as these date back to the earliest writings and are found in the theological writings of all people. It is widely stated and believed that "God has a plan for everyone" and things happen for a reason. These ideas were incorporated into modern science as the following quotations illustrate.

> We may regard the present state of the universe as the effect of its past and the cause of its future. An intellect which at a certain moment would know all forces that set nature in motion, and all positions of all items of which nature is composed, if this intellect were also vast enough to submit these data to analysis, it would embrace in a single formula the movements of the greatest bodies of the universe and those of the tiniest atom; for such an intellect nothing would be uncertain and the future just like the past would be present before its eyes.
>
> — Pierre Simon Laplace, *A Philosophical Essay on Probabilities*

As I have said so many times, God doesn't play dice with the world.

— Albert Einstein

Not surprisingly, these sentiments have found themselves into our literature.

What object is served by this circle of misery and violence and fear? It must tend to some end, or else our universe is ruled by chance, which is unthinkable.

— Sherlock Holmes, The Cardboard Box

Fortunately, for the human race some people decided to explore randomness and chance and see if there might be some "laws" or principles that govern it. Some of the earliest efforts to understand chance and randomness were inspired by gamblers. Girolamo Cardan (1501–1576) studied the questions associated with games of chance at the behest of gamblers. Other early contributors to the theory of probability were Fermat, Pascal and Descartes.

Another early effort to understand chance in the affairs of humans came when John Graunt (1620–1674) published *Natural and Political Observations Made Upon The Bills of Mortality* in London in 1662. He made the very surprising observation that while he could not tell who would die in a given year and how they would die, he could very accurately predict how many people would die in a "normal" year, and even how many people would die from what cause.

These observations from gambling and population studies fueled the development of probability and statistics and gave people some confidence that randomness followed some laws. The science of thermodynamics contributed significantly to the development of probability and statistics.

The birth of quantum mechanics in the twentieth century made randomness a central feature of science and caused Einstein to utter the famous quote presented earlier in this paper. Despite all efforts to remove randomness as a central feature of quantum mechanics, randomness appears to be at the center of the physical universe and to some extent, the Universe is ruled by chance.

If our understanding of quantum mechanics and of other physical theories is correct, then we should be able to get truly random numbers from a number of physical processes. Knuth [3, p. 3] describes some devices that were used to physically generate random numbers in the early days of computing. There are a number recent systems that generate random numbers based on physical phenomena. For example, Fourmilab [4] provides random bits derived from the radioactive decay of Cesium-137. Random.org [5] provides random numbers based on atmospheric phenomena. A joint effort between the Physics Department at Humboldt University (Germany) and PicoQuant GmBH [6] has produced the PQRNG 150, a quantum mechanical random number generating device that can be purchased [7]. One of the problems with standard physical devices is that they produce random bits very slowly. PicoQuant claims that the PQRNG 150 can produce more than 150 Mbits per second of random bits. Another option that might help to produce large quantities of truly random data is described in a paper by Gallego, Masasnes, De La Torre, Dhara, Aolita and Acfin [8].

3 Pseudo-Random Number Generators

From the very beginning of computing, there was a need for random numbers. Knuth [3, Chap. 3] provides a short history of the use of random numbers in computing and a lot of technical information about random numbers that should be more widely known. In 1946 John von Neumann first suggested using computer programs to generate "random numbers" [3, p. 3]. He also made the following statement about such programs.

> Any one who considers arithmetical methods of producing random digits is, of course, in a state of sin.
>
> — John von Neumann (1951) [3, p. 1]

As was noted from the beginning of modern computation, "random numbers" generated by some sort of deterministic computation cannot be truly random. They can, however, appear to be random. Such apparently random number sequences are referred to as *pseudo-random numbers* and the programs that generate them are known as *pseudo-random number generators* or *PRNGs*. To be truly useful, pseudo-random number sequences need to pass a wide variety of statistical tests. Knuth [3, Chap. 3] provides a good discussion of these tests and how they can be implemented.

As history has shown and this paper will document, people have made and continue to make serious mistakes in designing and running PRNGs. This inspired Robert Coveyou to write a paper in 1970 entitled "Random Number Generation Is Too Important to Be Left to Chance" [9]. In a similar vein, Knuth [3, p. 6] states that "...*random numbers should not be generated with a method chosen at random.*

We will now give a very general description of how PRNGs work. Generally, there is a finite set, S, of some sort. S is often a set of integers or real numbers. The PRNG can be thought of as a (deterministic) function $f : S \rightarrow S$. In general, S should be a large set and f should be a function that is easy to calculate, but whose inverse should be difficult to calculate. We start at some value a in S, called the *seed* and generate the sequence $a, f(a), f(f(a)), f(f(f(a))), \dots$. If we have done our work well, the sequence will appear to be random and pass whatever statistical tests we can devise.

Figure 1 shows the general structure of a PRNG. To understand this figure note that the sequence $a, f(a), f^2(a), f^3(a), \dots$, where $f^2(a) = f(f(a)), f^3(a) = f(f(f(a)))$, etc must repeat since S is finite. Suppose i is the smallest integer such that $f^i(a) = f^j(a)$ for some $i < j$. Note that the sequence $a, f(a), \dots, f^i(a)$ consists of distinct values. The value i is called the *index* of f. Pick the smallest value for j such that $i < j$ and $f^i(a) = f^j(a)$. The value $j - i$ is called the *period* of f and we will denote it by p. Note that $f^q(a) = f^{q+tp}(a)$ for all $q \geq i$ and all integers t. Generally, the function f is supplemented by a projection function π which converts the values $f^q(a)$ into whatever form is desired. For example, we might want to generate a pseudo-random sequences of 0s and 1s and π would convert the values $f^q(a)$ into 0s and 1s. Note that Figure 1 would remain the same regardless of whether

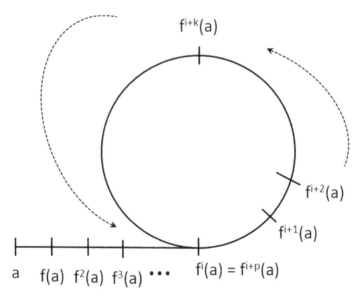

Figure 1   The Structure of a PRNG

f is an injective function or not. For some bounds on index and period, see Markowsky [10].

In general, the index is of little interest to PRNG designers and is often 0. Of great interest is the value of the period. In general, values for the PRNG are constrained to values that appear on the cycle. If the period is small, say less than 100,000, the PRNG is not very good for security purposes because an attacker can use a brute-force approach to compromising security. Just as 2 character passwords are not very secure because an attacker can try all of them, PRNGs with small periods are not very secure.

According to Wikipedia [11], the Mersenne Twister developed by Makoto Matsumoto and Takuji Nishimura [12] is currently the most used PRNG. It is based on the Mersenne prime $2^{19937} - 1$ and has a period of $2^{19937} - 1$. It passes most, but not all, randomness tests, but in its native form is not good for cybersecurity purposes in part because knowing 624 consecutive values permits one to figure out exactly where the PRNG is on its cycle and to generate all future outputs.

Besides Knuth's volume [3] which we have mentioned several times, we might also mention the article by Jerry Dwyer [13] that gives a more concise overview of PRNGs along with code examples. Wikipedia also provides some useful and current information about PRNGs.

4 RANDU

RANDU [14] is a PRNG that was introduced in the 1960s and used extensively for more than a decade. RANDU fails most randomness tests and is a bad PRNG. Knuth [3, p. 107] feels very strongly about its use as can be seen in the following quote where line 12 refers to a table containing the test results of RANDU and other PRNGs.

> Line 12 was, alas, the generator actually used on such machines in most of the world's scientific computing centers for more than a decade; its very name RANDU is enough to bring dismay into the eyes and stomachs of many computer scientists! ... the generator fails most three-dimensional criteria for randomness, and it should never have been used.

RANDU is an important example for us of a system failure. It was used by many people unfamiliar with the theory of PRNGs based on trust that a PRNG would not be used without careful thought and reasoning. It was used in all sorts of simulations and designs including those of nuclear weapons and reactors. One only hopes that RANDU's weaknesses did not lead to serious problems.

5 The Debian SSL Debacle

The discussion of the Debian SSL Debacle in this paper is based on papers by Ahmad [15], blog entries by Cox [16] and Schneier [17], the video by Bello and Bertacchini [18], and the video by Applebaum, Zovi and Nohl [19].

Ironically, the Debian SSL Debacle was caused by Debian developers trying to do a good thing. As is well known, memory errors are a fertile breeding ground for software failures of all types and often lead to breaches in cybersecurity. The Open Source community has developed tools to help find memory errors in software. In particular, "Valgrind" [20] is one such tool.

One of the developers running Valgrind on the Debian source code received a message warning about the use of an uninitialized variable in the function MD_Update in two places. Normally, this is an error in programming, but in this case the uninitialized variable was used together with other components to increase the amount of randomness in the OpenSSL module of the Debian operating system. In one location in the code, the use of MD_Update brought in other critically important sources of randomness in addition to the randomness imported from the uninitialized variable. Removing this instance of the MD_Update function critically damaged the OpenSSL module.

The OpenSSL module is used to encrypt communication for the operating system. For example, a person signing into a "secure" website is typically dependent on the OpenSSL module for providing good cryptographic strength. In general, the more randomness in a cryptographic system the better, so reducing the amount of randomness is a serious error.

The developer discussed with other Debian developers and also corresponded with OpenSSL developers. We shall examine the conversation in more detail in Section 10. He received an ambiguous reply that he interpreted as an approval to remove the lines in general, so he removed them from the program by commenting them out. It may be that the reply was just approving the removal of the lines in question just for debugging purposes, but the reply is poorly constructed and it is easy to see how the reply might have been interpreted as a blanket approval to remove those lines.

The main consequence of this action was to limit the PRNG so it would produce only 32,768 distinct values. While this is a large number for manual efforts, it is a relatively small number for a computerized "brute-force attack."

The lines were commented out in 2006, and it was not until 2008 that the error was discovered and the weakness of the resulting cryptographic system was established. Thus for about two years the cryptographic capabilities of Debian Linux were severely compromised. Hence the title of Ahmad's paper "Two Years of Broken Crypto" [15].

6 PRNG Problems in Android

Ironically, after the 2006 Debian SSL Debacle, Google committed the same sort of error in its PRNG for its Android operating system which is the most widely used operating system for mobile phones. The error again was not using all proper uses of entropy for the seed. The details are given in the following post to Google's Android Developers Blog [21] which appeared on August 14, 2013.

> The Android security team has been investigating the root cause of the compromise of a bitcoin transaction that led to the update of multiple Bitcoin applications on August 11.
>
> We have now determined that applications which use the Java Cryptography Architecture (JCA) for key generation, signing, or random number generation may not receive cryptographically strong values on Android devices due to improper initialization of the underlying PRNG. Applications that directly invoke the system-provided OpenSSL PRNG without explicit initialization on Android are also affected. Applications that establish TLS/SSL connections using the HttpClient and java.net classes are not affected as those classes do seed the OpenSSL PRNG with values from /dev/urandom.
>
> Developers who use JCA for key generation, signing or random number generation should update their applications to explicitly initialize the PRNG with entropy from /dev/urandom or /dev/random. A suggested implementation is provided at the end of this blog post. Also, developers should evaluate whether to regenerate cryptographic keys or other random values previously generated using JCA APIs such as SecureRandom, KeyGenerator, KeyPairGenerator, KeyAgreement, and Signature.
>
> In addition to this developer recommendation, Android has developed patches that ensure that Androids OpenSSL PRNG is initialized correctly. Those patches have been provided to OHA partners.

> We would like to thank Soo Hyeon Kim, Daewan Han of ETRI and Dong Hoon Lee of Korea University who notified Google about the improper initialization of OpenSSL PRNG.

7 The Bitcoin Compromise

As noted in Section 6, the Android PRNG was supposed to get a "random" seed from /dev/urandom, a protected system root file. However, the programmers did not reference this file in the code and left it to the user to pick a "random" seed. Of course, most users were not aware that they needed to do anything with the predictable result that a "random" seed was not picked in all cases.

So far there have been no verified consequences of the Debian OpenSSL Debacle, but we have a documented instance of some serious consequences of the Android failure. In particular, Goodin [22] describes the theft of $5,700 in bitcoins as a result of this flaw. It should be noted that the massive theft of $100 million in bitcoins [23] appears to be unrelated to this flaw and seems to be an instance of old fashioned, low tech fraud.

8 RSA, BSAFE and the NSA

The famous computer security firm RSA offers a cryptography library certified by NIST called BSAFE [24]. BSAFE can use a variety of PRNGs, but from 2004 to 2013 it made Dual_EC_DRBG the default PRNG for its library. This was a curious choice because as noted by Matthew Green [25]:

> Not only is Dual_EC hilariously slow – which has real performance implications – it was shown to be a just plain bad random number generator all the way back in 2006.

In particular, Microsoft Researchers Dan Shumow and Niels Ferguson at the Crypto 2007 rump session demonstrated that Dual_EC_DRBG might be compromised by the existence of a backdoor in this PRNG [26]. Shumow and Ferguson did not claim that a deliberate backdoor was placed in the PRNG, but warned that there could be one. Despite warnings from cryptographers about the dangers of using this PRNG, RSA continued to have use it as the default PRNG for BSAFE.

The Snowden leaks [24, 27, 28] showed that the concern about a backdoor were valid and that indeed such a backdoor was inserted into the PRNG by the NSA. In December 2013 Joseph Menn [29] revealed that the NSA paid RSA $10 million to make Dual EC DRBG the default PRNG for BSAFE.

This incident provides an example of the deliberate sabotage of a PRNG to obtain information. RSA is now left with the task of dealing with the fallout from this event. One embarrassing event was RSA warning its own customers about using the defaults in its own package [30]. Another embarrassing event was the withdrawal of a number of prominent cryptography experts from the RSA Annual Conference [31].

9 System Accidents

Perrow's book [1] contains detailed accounts of many industrial accidents as well as accidents in various other areas. Among the most fascinating accounts are those involving ships colliding in the ocean. A number of these collisions start off with the ships on courses that would not lead to a collision. Unfortunately, actions taken by the crew cause the collision. An example of such a collision is shown in Figure 2 which is derived from [1, p. 210]. Clearly, a collision is something both crews were eager to avoid, yet their actions led to a collision. Clearly, there is something about the marine transportation system that contributed to the collisions. There may, indeed, be failings on the part of the crew, but there are system failures that contributed to the accident.

Perrow uses the term system accident or normal accident throughout his book [1] but nowhere does he give a clear and concise definition. Furthermore, his focus is on physical systems and some of the factors he focuses on are less

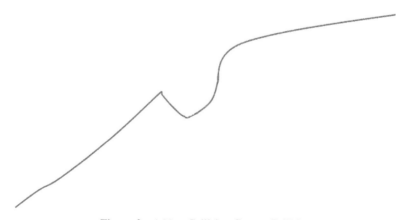

Figure 2   A Non-Collision Course Collision

relevant to virtual systems. We will review the appropriate terms from [1] and then propose how to interpret them for virtual systems.

First, Perrow spends a bit of time distinguishing between "incidents" and "accidents." Both of these "involve damage to a defined system that disrupts the ongoing or future output of that system" [1, p. 64]. Furthermore, Perrow divides systems into four levels: part, unit, subsystem and system. An incident is an event that occurs at the first two levels and an accident is an event that occurs at the last two levels. This leads Perrow to the follow formal definition [1, p. 66]. Note that the acronym ESF stands for engineered safety feature.

> We are ready for a formal definition. An *accident* is a failure in a subsystem, or the system as a whole, that damages more than one unit and in doing so disrupts the ongoing or future output of the system. An *incident* involves damage that is limited to parts or a unit, whether the failure disrupts the system or not. By disrupt we mean the output ceases or decreases to the extent that prompt repairs will be required. Since we have drawn a dividing line between the unit and the subsystem, and since many of the ESFs are clustered around that dividing line, it will often mean that an ESF will be one of the components that fails.

There are many features of the preceding definition that do not apply to the Debian OpenSSL Debacle. For example, the error that was introduced never interfered with the operation of the Debian operating system. In fact, the operating system ran without incident for two years before anyone noticed the error. In addition, it is not clear who was damaged by this bug and to what extent. We will address this issue further in Section 11. Clearly, the potential was there for massive cybersecurity breaches but we have no easy way to tell which security breaches were the result of this failure. In the case of physical systems and accidents like factory explosions and ship collisions it is often obvious what is damaged and to what extent. Of course, software systems can also cease to function as a result of errors or attacks, but it is not unusual for large software systems to have many errors and to function adequately. Of course, many of the errors are often minor and occur in very limited circumstances.

One of the major points that Perrow makes in [1] is that people, in particular system owners, like to find a scapegoat so that they are not obligated to fix the system. Often this is because system owners believe that it is cheaper to deal with the occasional mishap rather than fix the system [1, p. 67]. Perrow also has some very interesting points to make about the complexity of systems and the tight coupling of systems. There is no doubt that many software systems such as operating systems are extremely complex. Furthermore, there are few complex physical systems that do not depend on a complex software system for control.

In 2008 Perrow wrote an unpublished paper [32] dealing with software failures. While his analysis of technical issues leaves much to be desired, this paper is a fascinating collection of software failures of varying severity and is worth reading. The distinction between incident and accident is a valuable one even for virtual systems. We will use the term *incident* to describe a failure that has caused or is likely to be more of an inconvenience or nuisance than a serious failure. We will reserve the term *accident* for a failure that has caused or has the potential to cause serious damage or loss.

Perrow believes that serious accidents are often not just the result of one person's mistakes. Rather, they are often the result of a sequence of minor mistakes which combine to produce the serious accident. In this, they are aided by features of the relevant system that make the accident more likely. For example, in [1] Perrow shows how the operator is often blamed even though various safety devices did not work properly or gave misleading information. He spends lot of time in his analysis of the Three Mile Island nuclear accident [1, Chap. 1] showing how the operators were unable to get correct information from some of the gauges, yet they were blamed entirely for the accident.

We want to make the concept of a *booby trap* central to our definition of a system accident. A booby trap is designed to hurt someone who gets in its way. Booby traps can be set deliberately or unintentionally. Booby traps are often set by property owners to protect their property. Many polities have laws or regulations against the setting of booby traps because the final result of a booby trap is unpredictable and can be much worse than the consequence of the crime the booby trap is designed to prevent. Many people realize that booby traps such as wiring shotguns to doors are not intelligent things to construct. In fact, it is not unusual for property owners to be killed by their own booby traps [33].

For software systems we define a *booby trap* as some feature that makes it more likely that a user will make an error. One famous example of a software booby trap is the famous loss of the Mars Climate Orbiter spacecraft in 1999 [34]. The problem was that the two engineering teams that worked on the software for this system used different systems of measurement. In particular, one team used the metric system and the other team used the imperial system for measurements. As one would expect, at some point someone forgot to make the proper conversion and as a result a $125 million satellite was lost. It is clear that having two teams working with two different measurement systems is a booby trap. While one can try to blame "operator error" for the accident, it was clearly a system accident because the system was set up to make such an accident extremely likely. Despite this costly error, NASA continues to use both measurement systems for projects, although in 2007 it made the commitment to use the metric system for all operations on the lunar surface [35].

We propose the following definition: a software failure will be considered a *system accident* if it has serious consequences or the potential of having serious consequences and was caused by one or more booby traps. Our primary purpose is to identify booby traps in the software creation system, because even if they lead to minor errors in one situation, they can lead to more serious consequences in other situations.

10 The Debian SSL Debacle Revisited

In this section we will revisit the Debian OpenSSL Debacle from the point of view of finding booby traps in the open source development system that might be present in other systems as well. For convenient reference we will use DOD to refer to the Debian OpenSSL Debacle.

The first point to consider is the common lack of proper commenting in code. Many programmers either do not know how, don't have the time, or are too lazy write proper comments that really explain their reasoning and what the code does. They also tend not to point out booby traps in their code. This problem affects all types of software both proprietary and open source. This is *Booby Trap 1*. This booby trap was sprung in the DOD as illustrated by the fact that the developers who were fixing the "bug" did not properly understand the nature of the code they were working on.

Closely related to Booby Trap 1, is the fact that the programmers making the fix did not really understand the nature of what they were doing. To some extent this is *Booby Trap 2*, the lack of proper education. We view this as a system failure because the type of material found in Knuth [3, Chap. 3] is not generally taught in the computer science curriculum. Most computer science students have to take some sort of coursework in probability and statistics, but these courses typically do not talk about the problem of generating random numbers.

Closely related to the preceding booby traps, is the writing of overly clever code. This is code which does something in a very efficient and elegant manner and often combines multiple operations into one. Such code needs a lot of commenting and is often hard to understand for other programmers. It is often hard to understand for the programmer who wrote it once the creative insight passes. We call this *Booby Trap 3* and it was sprung in the DOD as discussed below.

While being proactive and finding errors in software before they manifest themselves is a good thing, there is a fundamental danger in using automated tools because these tools do not understand the logic of programs and perform their analysis at a very low level. Such tools can report errors, which are not truly errors. This is *Booby Trap 4*, and it was sprung in the DOD as seen from in the correspondence between Debian developers [36].

Valgrind has the property that once it finds an "error" it will track the effects of that error as they propagate through the entire software system. On the one hand this seems like a good idea, but in practice it leads to a large number of error messages. Experience shows that providing humans with too many warnings tends to make them ignore the warnings or even shut the system down. Perrow [1] describes many instances where safety systems were shut down before accidents because they were overwhelming the operators. This is *Booby Trap 5*. It was sprung in the DOD as illustrated in [36].

There is another booby trap associated with the use of powerful software testing tools. This is the potential to lead developers to venture into parts of the code that they do not understand well. This is *Booby Trap 6*. It was sprung in the DOD as illustrated in [36]. In this correspondence one of the developers notes that two lines in particular are the cause of the Valgrind error messages. He also notes that the function of these lines is to add "uninitialized numbers to the pool to create random numbers." Unfortunately, this developer did not fully understand that other sources of randomness were involved as well. The rest of the discussion in [36] turned into a technical discussion about using Valgrind, and the key point about contributing entropy was buried in the other discussion. One of the developers in the discussion was not completely comfortable with the discussion and realized that he was out of his depth in accessing the entropy issue and he contacted the OpenSSL developers [37]. In his note he made the statements shown in Figure 3.

There is another key point that needs to be made. The Debian developers identified two lines of code involved in generating the multitude of Valgrind errors. Both lines introduced an indeterminate amount of entropy from the uninitialized memory location, but one of the lines also introduced entropy from other sources such as system time, the PID, the UID and the random number generator. By focusing on the uninitialized variable, the developers overlooked all the other sources of randomness. Had these sources of entropy been introduced in separate lines, only the entropy coming from the uninitialized memory location would have been eliminated from the program. This is how Booby Trap 3 was sprung. The developers were led into this trap by inadequate commenting of the code which itself was too clever.

Figure 4 shows part of the reply 4 to the letter shown in Figure 3 is reproduced in Figure . It includes just one paragraph from the original letter which is emphasized. There are several problems with the reply. First, the reply begins with the phrase "Not much." It is not clear what "not much" refers to. If it refers to the sentence just above, then it suggests that the uninitialized variables do not add much entropy to the random number generator (RNG). This would support commenting out the lines in question. On the other hand, the original correspondence concluded with the line "What do you people think about removing those 2 lines of code?" It is possible that the writer was responding to that question and was stating that he did not think much of removing the two lines of code. Another ambiguity in the reply is the statement "If it helps with debugging, I'm in favor of removing them." Did the author mean that he was in favor of removing the lines but only for debugging purposes, or was he supporting the removal of the lines permanently? In any event, the Debian developer decided that this was the approval that he needed to remove the lines of code. The ambiguity in this communication was *Booby Trap 7*.

> What I currently see as best option is to actually comment out those 2 lines of code. But I have no idea what effect this really has on the RNG. The only effect I see is that the pool might receive less entropy. But on the other hand, I'm not even sure how much entropy some unitialised [sic] data has.
>
> What do you people think about removing those 2 lines of code?

Figure 3   Portion of a Letter to the OpenSSL Project

> *What I currently see as best option is to actually comment out those 2 lines of code. But I have no idea what effect this really has on the RNG. The only effect I see is that the pool might receive less entropy. But on the other hand, I'm not even sure how much entropy some unitialised [sic] data has.*
>
> Not much. If it helps with debugging, I'm in favor of removing them. (However the last time I checked, valgrind reported thousands of bogus error messages. Has that situation gotten better?)

Figure 4   A Reply to the Letter from Figure 3

11 Discovery, Mitigation and Consequences

Debian released Debian Security Advisory DSA-1571 [39] in 2008. It stated that Lucianon Bello discovered the flaw in the OpenSLL package used by Debian. It suggested some remediation such as regenerating various security keys. The announcement should have been enhanced to impress upon people the importance of this error. A variety of exploits aimed at this bug can be found in [18, 19, 39 – 42]. The sources just cited point to sources that should be consulted for additional exploits. We discuss an exploit in more detail in Section 12

Compromises affected not only Debian systems with the faulty software, but other systems that engaged in certain types of interactions with compromised computers. Reference [43] contains a discussion of weak keys, but this discussion would be very difficult for a non-expert to follow. Similarly, exploring DSA-1571 [39] provides many details, but again these would be difficult to follow for the non-expert. The failure to completely explain all consequences of this vulnerability and the failure to more widely alert the user community is *Booby Trap 8*.

There was another booby trap set by the Debian organization in the way that they handled the announcement of the vulnerability. In particular, they posted the vulnerability patch on May 7, 2008 but withheld the public announcement of the vulnerability until May 13, 2008. This is *Booby Trap 9*. The problem here is that there are skilled people who read code changes and who would understand the significance of this error even without the announcement. Not seeing the announcement at the same time as the patch, they would realize that there would be a window of opportunity to brute-force attack systems. Reference [19] reports that there was a sharp increase in the number of brute-force attacks against many hosts during the period between May 7 and May 13.

The inappropriate commenting out of two lines reduced the key space to a maximum of $2^{15} = 32,768$ keys. The total number of keys is greater because the set of 32,768 depends on the system used. In any event, generating the total number of keys for all systems is feasible with limited equipment. [19] even demonstrates how to use Amazon Cloud Services to generate the keys in a relative hurry for under $25.

References [15 – 19, 42] all discuss the consequences of this failure. Of particular interest is the graph in Figure 5 which is derived from [42]. It appears that after nearly six months more than 40% of the vulnerable certificates had not been updated. This is *Booby Trap 10*.

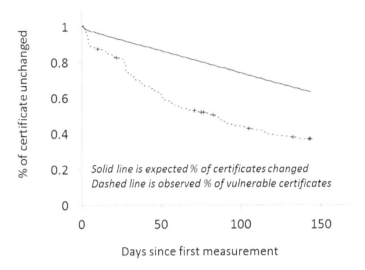

Figure 5 The Rate of Updating Vulnerable Certificates

12 An SSH Attack

In this section we wish to briefly describe in detail a possible attack against OpenSSH because of the weakness in OpenSSL. The reason OpenSSH was affected was because it uses the cryptography library in OpenSSL and hence uses the faulty PRNG from OpenSSL.

SSH supports a method for connecting to a server without using a password. The idea is based on the application of public key cryptography. In public key cryptography there are two keys that act as inverses of each other. Anything encoded using one key can be decoded using the other key. One key is called the *public key* and may be distributed without restriction. It can be published, displayed on a website or otherwise restricted. The other key is called the *private key* and must be maintained securely by the person wishing to receive messages encoded by the public key. In a successful public key system it should be extremely difficult (computationally impossible) to construct the private key given the public key.

Generally, the public and private keys are generated at the same time and are often based on a value supplied by a PRNG. If the PRNG is compromised and produces a small number of outputs, this would mean that only a small number of possible key pairs can be generated using that PRNG. Recall, that because of the reduced entropy to the PRNG in the Debian OpenSSL package, only 32,768 (2^{15}) key pairs are possible.

The passwordless login using SSH works as follows The person who wants to use such a feature generates a public-private key pair. The public key is stored on the server in a particular folder in the account space that belongs to the account which is to support passwordless login. Login then proceeds as follows. The user contacts the server and presents the login and the public key to the server. The server then checks whether the public key is present in the appropriate space in the account space and sends back a message encoded with the public key. In principle, only the holder of the matching private key can properly decode the message. Since the contents of the message are used as the basis of the link, the server has some confidence that the correct person has been connected.

With the Debian OpenSSL flaw only 32,768 key pairs were possible. This number of key pairs can be generated relatively easily to give all possible key pairs for key lengths of 1,024, 2,048 and 4,096 bits. The exploit proceeds as follows. The attacker picks a user ID of someone likely to use passwordless login such as a system administrator. Passwordless login is used by many system administrators because they want to quickly log into many different computers. The attacker then presents the user ID along with each of the public keys in turn until the server gives a proper response at which time the attacker uses the corresponding private key to initiate the session. While 32,768 choices seems like a lot, this is a not a large number of trials for a computerized brute force attack.

TJ O'Connor's book, *Violent Python* [45, pp. 41–55] describes such an attack in detail and provides Python code for this example. Details on SSH are available in the book by Barrett, Silverman and Byrnes [46]. A detector of weak key material is available at the Debian.org site [39] and directly at [47]. This detector is written in Perl and contains information about the keys.

The author would like to thank his RIT colleagues Bruce Hartpence and Bill Stackpole for some useful discussions about SSH.

13 Conclusions and Suggestions for Improvement

In this paper we identified ten booby traps that led to the DOD and contributed to worsening the consequences. We list them below with a brief description and some suggestions for dealing with each booby trap. It is clear that these booby traps are not unique to the DOD and perhaps the lessons learned here can be helpful elsewhere.

1. *Poor Commenting.* This is a standard problem in coding and it is not clear that we are making progress. Figure 6 [44] shows a more recent version of the code that was involved in the DOD. Note that while there is a comment warning people not to remove a particular instance of MD_Update, there is no explanation of why an uninitialized variable is being used. Furthermore, there are several other references to MD_Update in the same code section but there is no explanation of what these other calls are achieving. Poor commenting is a system failure. It is not properly taught in most computer science programs and there are few good tools for writing useful comments. Knuth [48, 49] has put forward a philosophy of programming and produced tools to support his approach, but so far this effort has had limited impact. More effort needs to be put into this initiative and similar initiatives. On a related note, more widespread use of *Test-Driven Development* [50, 51] can provide programmers with useful information about what code is supposed to do along with tests that can help limit errors stemming from code modification.

2. *Lack of Proper Instruction.* As noted in this paper, most computer science professionals have an inadequate basis for understanding the difficulties of using PRNGs. We suggest that material dealing with random numbers and PRNGs become part of the standard computer science curriculum.

```
265        MD_Init(&m);
266        MD_Update(&m,local_md,MD_DIGEST_LENGTH);
267        k=(st_idx+j)-STATE_SIZE;
268        if (k > 0)
269                {
270                MD_Update(&m,&(state[st_idx]),j-k);
271                MD_Update(&m,&(state[0]),k);
272                }
273        else
274                MD_Update(&m,&(state[st_idx]),j);
275
276        /* DO NOT REMOVE THE FOLLOWING CALL TO MD_Update()! */
277        MD_Update(&m,buf,j);
278        /* We know that line may cause programs such as
279            purify and valgrind to complain about use of
280            uninitialized data.  The problem is not, it's
281            with the caller.  Removing that line will make
282            sure you get really bad randomness and thereby
283            other problems such as very insecure keys. */
284
285        MD_Update(&m,(unsigned char *)&(md_c[0]),sizeof(md_c));
```

Figure 6   More Recent Version of the OpenSSL Source Code

3. *Overly Clever Coding.* Programmers should realize that while very clever coding might save some space and coding time, it is very difficult to understand and is a fertile breeding ground of errors. Figure 6 [44] shows there there are now more calls to MD_Update so the functionality has been separated to some extent, but no information is given about what each call is doing.

4. *Uncritical Use of Automated Software Analysis Tools.* Figure 7 which comes from [18] shows the comparison of code before and after the modification. The right half of the figure has the changes in green. The presentation is a bit deceptive because the effect of the introduced lines is to remove the lines dealing with MD_Update from the program. We need to make it more obvious to an observer that the changes have affected the lines containing MD_Update.

5. *Overwhelming Error Messages.* Thought needs to be given on how to demonstrate all weaknesses in some code without overwhelming the person using the automated tool.

6. *Repairs by Nonexperts.* If the Federal government wants to help improve US cybersecurity it should consider offering code reviews for critical software. Clearly, it is not reasonable for the Federal government to review all code, but Debian and its derivatives such as Ubuntu are very popular and are the basis of a significant number of servers on the Internet. Given the recent Snowden revelations and the deliberate introduction of weaknesses into the BSAFE package by the NSA, it is not clear that one can depend on the Federal government to give a honest appraisal.

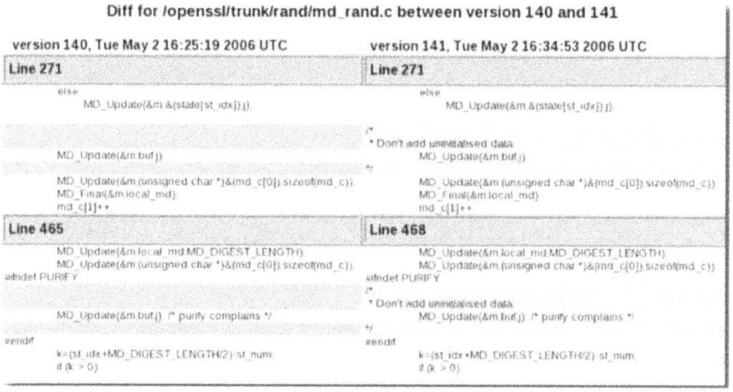

Figure 7   Diff Obscuring High Level Understanding

7. *Ambiguous Communication.* Perhaps a more formal process can be instituted here to make sure that all questions are posed and answered unambiguously. It appears that the standard method of just mailing to a site and having people selectively reply to sections of the e-mail is fraught with danger.

8. *Poorly Distributed and Overly Technical Announcements.* We need to get more people with public relations and communications skills active in the open source community. These people must be made to feel welcome and not put down by the technical community since they have an important job to perform.

9. *Posting Patches Prematurely.* Clearly, it is recommended that organizations not publicly post patches before they announce vulnerabilities. Perhaps they can post patches only to vetted customers to give them a chance to update their systems.

10. *User Community Not Taking Cybersecurity Seriously Enough or Perhaps not Having the Resources to Deal With Critical Issues.* This is a challenging problem that deserves a separate discussion.

We wish to conclude with one additional recommendation. While systems such as Knuth's CWEBB and Beck's Test-Driven Development introduce more formal procedures into code design and creation, there is still room for global formal methods. Books by Holtzmann [52] and Berg, Boebert, Franta and Moher [53] provide a good place to acquire the basics. A current open-source software tool is called Spin and and programmers are encouraged to become become familiar with it [54, 55].

References

[1] Charles Perrow, *Normal Accidents*, Princeton University Press, 1999.
[2] George Markowsky, "Was the 2006 Debian SSL Debacle a System Accient?,"
[3] Donal Knuth, *The Art of Computer Programming*, 3rd ed., 1998, Vol. 2, *Seminumerical Algorithms*.
[4] Geiger counter-based random number generator. http://www.fourmilab.ch/hotbits/.
[5] An atmospherically based random number generator. http://www.random.org/.
[6] Quantum Mechanics Random Number Generator Project, 'http://qrng.physik.hu- berlin.de/.

[7] PQRNG 150 (Quantum Mechanics Random Number Generator) Pico-Quant GmBH, http://www.picoquant.com/products/category/ quantum-random-number-generator/pqrng-150-quantum-random-number-generator.

[8] Rodrigo Gallego, Lluis Masanes, Gonzalo De La Torre, Chirag Dhara, Leandro Aolita and Antonio Acfin, "Full Randomness from Arbitrarily Deterministic Events," *Nature Communications* 4, Article number: 2654, http://conference.iiis.tsinghua.edu.cn/QIP2013/wp-content/uploads/2012/12/qip2013_submission_7.pdf

[9] Robert Coveyou, "Random Number Generation Is Too Important to Be Left to Chance," *Studies in Applied Mathematics*, III, 1970, pp. 70–111.

[10] George Markowsky, "Bounds on the Index and Period of a Binary Relation on a Finite Set,"*Semigroup Forum*, v. 13, 1877, pp. 253–259.

[11] "Mersenne Twister," *Wikipedia*, February 1, 2014, http://en.wikipedia.org/wiki/Mersenne_twister.

[12] Makoto Matsumoto and Takuji Nishimura, "Mersenne Twister: A 623-Dimensionally Equidistributed Uniform Pseudo-Random Number Generator," *ACM Trans. on Modeling and Computer Simulation*, v. 8, n, 1, January, 1998, pp. 3–30.

[13] Jerry Dwyer, 'Quick and Portable Random Number Generators," *Dr. Dobb's Journal*, June 1, 1995, http://www.drdobbs.com/quick-and-portable-random-number- generat/184403024

[14] "RANDU," *Wikipedia*, February 2, 2014, http://en.wikipedia.org/wiki/RANDU

[15] David Ahmad, "Two Years of Broken Crypto," *IEEE Security & Privacy*, 2008, pp. 70–73.

[16] Russ Cox, "Lessons from the Debian/OpenSSL Fiasco,", blog, http://research. swtch.com/openssl.

[17] Bruce Schneier, "Random Number Bug in Debian Linux," *blog Schneier on Security*, http://www.schneier.com/blog/archives/2008/05/random_number_b.html

[18] Luciano Bello and Maximiliano Bertacchini, "Predicatable PRNG In the Vulnerable Debian OpenSSL Package: The What and the How," DEFCON 16, August 8–10, 2008, Las Vegas, NV, http://www.youtube.com/watch?v=yXr7KBC3G3I. The slides are available at http://www.citedef.gob.ar/si6/descargas/openssl-debian-defcon16 .pdf.

[19] Jacob Applebaum, Dino Dai Zovi, and Karsten Nohl, "Crippling Crypto: The Debian OpenSSL Debacle,", http://www.youtube.com/watch?v=Qd knzkoN_aI. The slides are available at http://trailofbits.files.wordpress. com/2008/07/hope-08- openssl.pdf.

[20] Valgrind Website, http://valgrind.org/.

[21] "Some SecureRandom Thoughts," *Android Developers* Blog, August 14, 2013, http://android-developers.blogspot.com/2013/08/some-securerand om- thoughts.html.

[22] Dan Goodin, "Google Confirms Critical Android Crypto Flaw Used in $5,700 Bitcoin heist,", em Ars Technia, August 14, 2013, http://arstechnica.com/security/ 2013/08/google-confirms-critical-andro id-crypto-flaw-used-in-5700- bitcoin-heist/.

[23] Jim Edwards, "A Thief Is Attempting To Hide $100 Million In Stolen Bitcoins And You Can Watch It Live Right Now," *Business Insider*, December 3, 2013, http://www.businessinsider.com/a-thief-is-attempting-to-hide-100- million-in-stolen-bitcoins-and-you-can-watch-it-live-right-now-2013–12.

[24] "RSA BSAFE," *Wikipedia*, February 2, 2014, http://en.wikipedia.org/ wiki/ RSA_BSAFE.

[25] Matthew Green, "RSA Warns Developers Not to Use RSA Products," blog, http://blog.cryptographyengineering.com/2013/09/rsa-warns- developers-against-its-own.html.

[26] Dan Shumow and Niels Feguson, "On the Possibility of a Back Door in the NIST SP800-90 Dual Ec Prng," http://rump2007.cr.yp.to/15-shumow.pdf.

[27] Glenn Greenwald, "Edward Snowden: the whistleblower behind the NSA surveillance revelations," *The Guardian*, June 9, 2013, http://www.theguardian.com/world/2013/jun/09/edward-snowden-nsa-whistleblower-surveillance.

[28] "Edwward Snowden," *Wikipedia*, February 2, 2014, http://en.wikipedia. org/wiki/Edward_Snowden#cite_note-cnn-hotel-104.

[29] Joseph Menn, "Exclusive: Secret Contract Tied NSA and Security Industry Pioneer," *Reuters*, December 20, 2013, http://www.reuters.com/ article/2013/12/20/ us-usa-security-rsa-idUSBRE9BJ1C220131220.

[30] Dan Goodin, "Stop using *NSA-influenced* code in our prod-ucts, RSA tells customers," *Ars Technica*, September 19, 2013, http://arstechnica.com/security/2013/09/ stop-using-nsa-influence-code-in-our-product-rsa-tells-customers/

[31] Dan Goodin, "More Researchers Join RSA Conference Boycott to Protest $10 million NSA Deal," *Ars Technia*, January 7, 2014, http://arstechnica.com/security/2014/01/more-researchers-join-rsa-conference-boycott-to-protest-10-million-nsa-deal/

[32] Charles Perrow, "Software Failures, Security, and Cyberattacks," Online article at http://www.cl.cam.ac.uk/~rja14/shb08/perrow.pdf, http://www.tatup-journal.de/english/tatup113_perr11a.php.

[33] "Booby-Trapped Onion Patch Kills Owner, Son,", Los Angeles Times, June 12, 1994, http://articles.latimes.com/1994–06-12/news/mn-3270_1_onion-patch.

[34] "NASA's metric confusion caused Mars orbiter loss," CNN, September 30, 1999, http://www.cnn.com/TECH/space/ 9909/30/mars.metric/.

[35] "Metric Moon,", NASA Science News, online article, http://science1.nasa.gov/science-news/science-at-nasa/2007/08jan_metricmoon/.

[36] Richard Kettlewell, "valgrind-clean the RNG," Debian Bug report logs -#36516, http://bugs.debian.org/cgi-bin/bugreport.cgi?bug=363516.

[37] Kurt Roeckx, "Random number generator, uninitialized data and valgrind,", OpenSSL Mail Archive, http://www.mail-archive.com/ openssl-dev@openssl. org/msg21156.html.

[38] Ulf Möller, "Random number generator, uninitialized data and valgrind,", OpenSSL Mail Archive, http://www.mail-archive.com/ openssl-dev@openssl.org/ msg21157.html.

[39] Debian Security Advisory 1571, http://www.debian.org/security/2008/ dsa-1571.

[40] Luciano Bello, "Exploiting DSA-1571: How to break PFS in SSL with EDH," http://www.lucianobello.com.ar/exploiting_DSA-1571/

[41] Ben Feinstein, "Loaded Dice: SSH Key Exchange & the Debian OpenSSL PRNG Vul- nerability," ToorCon X, September 27, 2008. Slides are available at http://www.cs.unh.edu/it666/reading_list/Keys/ ssh_key_exchange.pdf.

[42] Scott Yilek, Eric Rescorla, Hovav Shacham, Brandon Enright, and Stefan Savage, "When Private Keys are Public: Results for the 2008 Debian OpenSSL Vulnerability", *ICM'09*, November 4–6, 2009, Chicago Illinois, USA.

[43] "The SSL Keys and Various Applications," http://wiki.debian.org/ SSLkeys

[44] OpenSSL Source Code for md rand.c, https://github.com/luvit/openssl/ blob/master/openssl/crypto/rand/md_rand.c

[45] TJ. O'Connor, *Violent Python: A Cookbook for Hackers, Forensic Analysts, Penetration Testers and Security Engineers*, Syngress, Waltham, MA, 2013.

[46] Daniel J. Barrett, Richard E. Silverman, and Robert G. Byrnes, *SSH, The Secure Shell, The Definitive Guide*, 2nd ed., O'Reilly, Sebastopal, CA, 2005.

[47] "A Detector for Known Weak Key Material,", *Debian.org*, http://security.debian.org/project/extra/dowkd/dowkd.pl.gz.

[48] Donald E. Knuth, *Literate Programming*, Center for the Study of Language and Information, 1992.

[49] Donald E. Knuth and Silvio Levy, *The CWEB System of Structured Documentation*, Version 3.6, Addison-Wesley, 1994, http://sunburn.stanford.edu/knuth/cweb. html.

[50] Kent Beck, *Test-Driven Development by Example*, Addison-Wesley, 2002.

[51] "Test-Driven Development," *Wikipedia*, February 4, 2014, http://en.wikipedia.org/wiki/Test-driven_development.

[52] Gerard j. Holzmann, *Design and Validation of Computer Protocols*, Prentice Hall, 1991.

[53] H. K. Berg, W. E. Boebert, W. R. Franta and T. G. Moher, *Formal Methods of Program Verification and Specification*, Prentice Hall 1982.

[54] "Home Page for Spin," http://spinroot.com/spin/whatispin.html.

[55] Gerard J. Holzmann, *The SPIN Model Checker: Primer and Reference Manual*, Addison- Wesley, 2003.

Biography

George Markowsky spent ten years at the IBM Thomas J. Watson Research Center where he served as Research Staff Member, Technical Assistant to the Director of the Computer Science Department, and Manager of Special Projects. He came to the University of Maine as the first Chair of the Computer

Science Department. During 2004–2005 he was Dean of the American-Ukrainian Faculty at Ternopil National Economic University in Ukraine. In 2006–2007 he was Visiting Professor at Rensselaer Polytechnic Institute Lally School of Management and Technology. He became Chair of the Computer Science Department again in 2008 and served in that capacity until the Department became part of the new School of Computing and Information Science at which time he became the Associate Director of the School. In 2013–2014 he has been a Visiting Scholar in the Department of Computing Security at the Rochester Institute of Technology. He is currently Professor of Computer Science at the University of Maine. George Markowsky has published 113 journal papers, book chapter, book reviews and conference papers on various aspects of Computer Science and Mathematics. He has written or edited 15 books and reports on various aspects of computing. He also holds a patent in the area of Universal Hashing. His interests range from pure mathematics to the application of mathematics and computer science to biological problems. He has also built voice controlled and enhanced keyboard terminals for use by paralyzed individuals. He is very active in homeland security and is the director of the University of Maine Homeland Security Lab.

Fast Network Attack Modeling and Security Evaluation based on Attack Graphs

Igor Kotenko and Andrey Chechulin

Laboratory of Computer Security Problems, St. Petersburg Institute for Informatics and Automation of the Russian Academy of Sciences (SPIIRAS), 39, 14th Liniya, St. Petersburg, Russia
{ivkote, chechulin}@comsec.spb.ru

Received 3 February 2014; Accepted 27 April 2014;
Publication 2 June 2014

Abstract

The paper suggests an approach to network attack modeling and security evaluation which is realized in advanced Security Information and Event Management (SIEM) systems. It is based on modeling of computer network and malefactors' behaviors, building attack graphs, processing current alerts for real-time adjusting of particular attack graphs, calculating different security metrics and providing security assessment procedures. The novelty of the proposed approach is the use of special algorithms for construction, modification and analysis of attack graphs aimed at rapid security evaluation. This allows using this approach in SIEM systems that operate in near-real time. The generalized architecture of the Attack Modeling and Security Evaluation Component (AMSEC), as one of the main analytical components of SIEM systems, is outlined. The main components and techniques for attack modeling and security evaluation are defined. A prototype of the AMSEC is presented. Experiments with this prototype are evaluated.

Keywords: network attack modeling, attack graphs, security evaluation, near real time, security information and event management.

Journal of Cyber Security, Vol. 3 No. 1 , 27–46.
doi: 10.13052/jcsm2245-1439.312

1 Introduction

In SIEM systems the security administrator should check whether network configuration parameters and security procedures provide the necessary security level. Moreover, at exploitation stage, a lot of security events and alerts have place, the configuration of computer networks can be changed, new vulnerabilities may be discovered, new attack exploits can be developed, new services are able to be added, etc. That's why it is necessary continually to perform network monitoring, analyze available vulnerabilities, current security events and evaluate security level.

The main purpose of the presented work was to increase the speed of network security evaluation. To achieve this purpose an original set of models, algorithms and techniques was developed. The performance of the proposed technique, as it was demonstrated in experiments, allows using the results of this research in systems operating in near real time.

Key elements of suggested architectural solutions for network attack modeling and security evaluation in SIEM systems are using a comprehensive security repository, effective attack graph (tree) generation techniques, taking into account known and new attacks based on zero-day vulnerabilities, stochastic analytical modeling, and interactive decision support to choose preferred security solutions [12, 14–16].

This paper considers the state-of-the-art in network attack modeling and security evaluation based on attack graphs, the essence of the approach to analytical attack modeling as well as a generalized architecture of Attack Modeling and Security Evaluation Component (AMSEC) suggested in the EU MASSIF project [25].

The paper is structured as follows. Section 2 summarizes related work. The common framework for computer attack modeling and security assessment is presented in Section 3. A technique for attack graph construction and analysis is outlined in Section 4. Implementation issues are outlined in Section 5. Section 6 describes some experiments that we have carried out. The Conclusion summarizes the main results and suggests some future work.

2 Related Work

One of the fist descriptions of attack graphs was suggested by Schneier [4]. In this work the approach to manual construction of attack graphs was used for security evaluation. Each graph contains nodes that represent attacker's aims and nodes that represents attack actions.

Moore et al. [1] proposed a structured and reusable tree-based form for attacks description and modeling. Swiler and Phillips [20] presented one of the first software tools for attack graph generation. Each node of attack graph modeled in this tool represents an attack state and edges specify the attacker's actions.

Lippmann and Ingols [33] considered a tool that is used to construct and analyze automatically attack graphs for detection of firewall configuration defects and host critical vulnerabilities. Information about network vulnerabilities is collected by Nessus security scanner [30], and this information must be manually entered in the database.

Ingols et al. [18] extended this approach to take into account modern network attacks and countermeasures. Particularly, they suggest the improvements to model additional modern threats and countermeasures.

The list of papers in this area is very huge.

In addition to theoretical work, there are several software implementations of security evaluation systems based on different principles of security analysis. For example, Gamal et al [23] presented a security analysis system OpenSKE (Open Security Knowledge Engineered), which uses an expert system to assess security. The other example is CAULDRON [6], developed at George Mason University, which is also based on the construction and analysis of attack graphs.

The most common platform and vulnerability specifications are proposed by MITRE Corporation [27]: CPE[7], CVE[8], CVSS[9] and CAPEC [5].

Common Platform Enumeration (CPE) [7] provides a unified description language for information technology systems, platforms, and packages. It is based on the generic syntax for Uniform Resource Identifiers (URI). CPE contains a formal name format, a language for specifying complex platforms, a method for checking names against a system, and a description format for binding text and tests to a name.

Common Vulnerabilities and Exposures (CVE) [8] dictionary contains the list of known information security vulnerabilities and exposures. Each vulnerability/exposure has a unique identifier. This enables data exchange between different security products and gives an opportunity to evaluate different tools and services.

Common Vulnerability Scoring System (CVSS) [9] is an open and standardized vulnerability scoring system. CVSS gives an opportunity to prioritize and coordinate a reasonable response to security vulnerabilities via the base, temporal and environmental properties of vulnerability.

Usage of the National Vulnerability Database (NVD) [29] based on the CVE dictionary is the basis for constructing of attack graph via known vulnerabilities and Common Attack Pattern Enumeration and Classification (CAPEC) [5] provides data about attack patterns.

Important aspects of building a data repository, allowing promptly provide data for the security evaluation system based on attack graphs, are presented in [17].

Kheir et al. [28] propose to extend the use of CVSS metrics in the context of intrusion response, by supplying dynamic information about system configuration and service dependencies structured within dependency graphs.

Security metrics are an important element of the security evaluation system. From the system security level point of view a set of security metrics can be outlined: integral metrics of the common security level of system, metrics that define topological characteristics, malefactor characteristics and attack characteristics [3, 19, 32, 34, 35, 37].

The analysis of network security against unknown zero day attacks is also an important related topic of research [10, 21, 22, 24].

3 Common Framework

The Attack Modeling and Security Evaluation Component (AMSEC) is intended to complement the SIEM analysis functionality with the capability of network attack modeling and security evaluation [13, 14, 16].
The *main inputs* for AMSEC are:

- configuration of the computer network (system);
- policy determining a set of permissions or policy rules;
- event and alerts;
- external databases (DBs) of vulnerabilities, attacks, platform, etc.;
- possible malefactor profiles (as a set of malefactor characteristics);
- required values of security metrics (as a set of requirements to security).

The *main results* of AMSEC are as follows:

- vulnerabilities detected;
- possible routes (graphs) of attacks and attack goals;
- payload internal dependencies;
- bottlenecks ("weak places") in network security;
- preliminary attack graphs;
- adjusted attack graphs based on changes in the network and alerts;

- predictions of the malefactor's next steps taking into account the current situation;
- security metrics, which can be used for general security level evaluation of computer network (system) and its components;
- attack and countermeasures impacts;
- guidelines for increasing the security level and solutions based on security measures/policies/tools.

The *general architecture* of the AMSEC and its interaction with other components of SIEM system are shown in Figure 1. Connections, depicted in the figure, show the direction of interactions between different components.

Data repository updater downloads the open databases of vulnerabilities, attacks, configurations, weaknesses, platforms, and countermeasures from the external environment (sending requests to external databases for updates and communicating with data sources).

Specification generator converts the information about network events, configuration and security policy, from other SIEM components or from users, into an internal representation.

Malefactor modeler determines malefactors' individual characteristics, skill level, their initial position (insider/outsider, available points of entry,

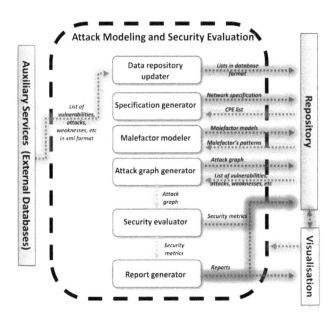

Figure 1   General architecture of AMSEC

etc.), the set of permissions, possible actions/attacks already fulfilled (which can be predicted according to events and alerts) and knowledge about the analyzed network.

It also recognizes the most probable malefactor model based on detected attacks and modifies the attack graph based on the changes of the network.

Attack graph generator builds attack graphs by modeling sequences of malefactor's attack actions in the analyzed computer network using information about available attack actions of different types, services dependencies, network configuration and used security policy. Attack graph generator can also build attack traces taking into account zero-day vulnerabilities – unknown vulnerabilities which are required to compromise network assets.

Security evaluator assists the selection of solutions (validated events and alerts, possible future security events, countermeasures) needed for other SIEM components. It simulates stochastically multi-step attacks and studies the cost and effect of various countermeasures. For example, it generates combined objects and calculates their security metrics in order to evaluate the common security level and possibly make recommendations on strengthening it.

Report generator shows vulnerabilities detected by the AMSEC, repre-sents "weak" places, generates recommendations on strengthening the security level and depicts other relevant security information.

The AMSEC operates in two main modes (Figure 2):

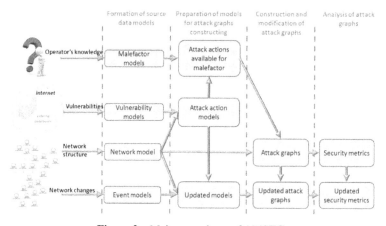

Figure 2   Main procedures of AMSEC

- Design time (or configuration), when the AMSEC is used for design and initial analysis of the network analyzed (or the system under protection). It is a non real-time mode;
- Exploitation, when the AMSEC is used for real-time or near real-time operation of the SIEM system.

Each of the modes may in turn be divided into the following steps (Figure 2):

- Creating initial data models;
- Preparation of the models for construction of attack graphs;
- Generation of attack graphs;
- Analysis of attack graphs.

The functionality of the AMSEC requires the presence of the vulnerability database loaded from the Internet as well as repository which will store the input and output data of the AMSEC. As external repository for the AMSEC the Common Repository is used.

The second component with which AMSEC has tightly integration is the Visualization Component.

Let us consider the main procedures of AMSEC and the interaction of the AMSEC with SIEM components on main modes.

Design (configuration) stage.

The AMSEC needs to have a detailed description of protected network topology and configuration for correct and efficient operation. This information is retrieved from the user (through the Visualization system), from predefined data (through Repository) and from sensors placed in the network. As a result, the AMSEC produces attack graphs and calculates security metrics.

Attack graphs can be used to refine event processing rules, and security metrics can be used for decision support and reaction to form the list of recommendations to increase the security level. Since at this stage real-time mode is not required, the information flow can go through the Repository.

Exploitation stage.

There are several tasks performed by the AMSEC at this stage: attack graphs adjustment; attack detection improvement by searching matches between real-time events and attack graphs; security metrics evaluation and prediction of potential threats and attacks.

4 Attack Graph Construction and Analysis Technique

We developed a set of algorithms of attack graphs construction, modification and analysis that can be implemented in various stages of a network lifecycle. The main steps of these algorithms are shown in Figure 3.

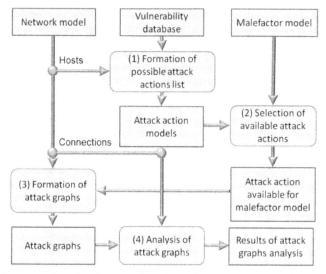

Figure 3   Representation of main steps of attack graph construction, modification and analysis algorithms

Blocks (1), (2) and (3) perform various tasks in the attack graphs construction and modification processes for the design and operation stages of the network lifecycle. Block (4) performs the analysis of attack graphs.

Let us consider the application of the algorithms for network lifecycle in more detail.

On the design stage the algorithms for *construction and analysis* of attack graphs are used. Blocks 1 – 3 perform the following steps to construct attack graph:

1. The Block 1 generates a list of possible attack actions divided into groups according to the following parameters: attack class, required access and required knowledge level of the malefactor. For each group, in turn, a list of realizable specific attack actions and vulnerabilities is created.
2. For each network's host, Block 2 selects the attack actions that can be used by each malefactor model.
3. On the basis of network host connections and available attack actions, Block 3 generates host accessibility connections taking into account available attack actions. These connections are constructed for all selected malefactor models.
4. Based on accessibility connections, the Block 3 also forms the attack graphs for the initial access points for all malefactors.

On the exploitation stage the computer network represents a continuously changing object, i.e. its structure and the elements (e.g., hosts) may be changed over time. The computer network model and consequently attack models and their evaluation results also can be changed according with the changes in the real network. An important peculiarity of this stage is that the most of existing attack modeling systems may require a lot of time and resources for this kind of modeling more or less equal with the design stage.

On the exploitation stage the proposed *modification algorithm* uses different blocks according to a network's change type. The possible network change types are as follows:

- change of network topology (adding or removing connections between network hosts);
- change in the hosts (adding, removing or changing of software and hardware, policies, etc.);
- adding, removing or changing of malefactor models;
- adding, removing or changing vulnerabilities.

For each type of change a subalgorithm is developed. It allows to minimize the time required for model updating.

There are three classes of changes grouped by impact on attack graphs:

1. that do not affect the attack graph;
2. that reduce the attack graph (for example, uninstalling of some software or hardware);
3. that extend the attack graph (for example, installing of some vulnerable software or hardware).

Similarly, the changes are grouped by affecting to the available malefactor's attack actions. This approach allows us to significantly reduce the models' modification time.

A security evaluation often requires detailed analysis of all elements of the network model. Such an analysis can take a long time. Thus, the exact values of security metrics (for example, attack impact or common security level) may not be available after the beginning of the evaluation until some time.

To solve this problem, we propose the following approach: the security evaluation problem is represented as a series of algorithms with varying computational complexity. An example of block diagram illustrating this approach is shown in Figure 4. For this purpose, first of all, the structure and content of input data are changed.

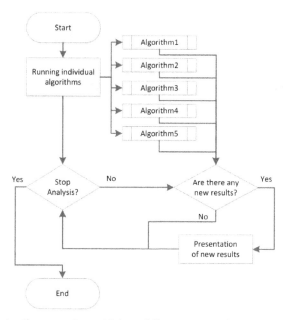

Figure 4 Block diagram of combining different approaches to security evaluation Implementation

The algorithms can use (in the order of increasing the complexity of the analyzed network model):

1. a list of individual hosts, excluding topology;
2. a simplified model of the network topology (subnets are grouped by the criticality level and presented in the form of integrated models);
3. a complete network model including separate models for each object in the network.

For instance, for security evaluation based on the network model (1) the vulnerabilities of hosts are taken into account. In this case the network is considered insecure if at least one vulnerability with high-risk exists in the network. This approach allows us to define a secure network as a network that has no vulnerabilities. But if vulnerabilities on some hosts exist, the result of security evaluation is inaccurate, since the malefactor might not have access to these hosts.

Security evaluation based on a simplified topology (2) uses an approach similar to (1), but only for certain sub-networks. Thus, the attack graph is formed in a simplified form and comprises substantially fewer possible routes.

This approach allows a more accurate security evaluation, but at the level of sub-networks has the same disadvantages as the first approach.

Security Evaluation (3), conducted on the basis of a complete network model, is the most accurate, but requires a much more time for analysis, since the attack graph as a rule will include a great number of attack routes.

5 Implementation

The prototype of Attack Modeling and Security Evaluation Component (AMSEC) is designed as an element of SIEM system. The prototype of AMSEC includes the following functional subsystems: data storage subsystem; generator of network and malefactor models; generator of attack graphs (working in the mode of construction and modification); data analysis subsystem [13, 14, 16].

The generator of random network models was implemented as a data source for the experiments. It allows shaping the input data, containing descriptions of interconnected hosts. The resulting models may contain the specifications of user and server (databases, application servers, etc.) hosts with different parameters of the stored information criticality.

To obtain data on real computer networks, we used the security scanner MaxPatrol [26]. This scanner has two main scanning modes: system (Audit) and network (PenTest). System scanning mode assumes that the special agent is installed on each host. This agent allows gathering all necessary information about the software and hardware of the host and possible vulnerabilities in the operating system settings and individual programs. Network scanning mode allows seeing the network through the eyes of the malefactor. In this mode, the scanner implements scanning from an external host and collects information available outside about each network host.

Information obtained in the system scanning mode:

- list of software (applications) and operating system components detected;
- list of vulnerabilities found in software (applications) and components of the operating system.

Information obtained in the network scanning mode:

- list of network services and protocols found;
- list of vulnerabilities found in network services and protocols.

The main standards used for processing of initial data are depicted in Table 1.

Table 1 Initial data representation formats

Data	Format
Data about software and hardware of hosts	CPE (Common Platform Enumeration) [7]
Network topology	XML
Vulnerabilities	CVE (Common Vulnerabilities and Exposures) [8]
Malefactor model	XML
Data on attacks	CAPEC (Common Attack Pattern Enumeration and Classification) [5]
Security metrics	CVSS (Common Vulnerability Scoring System) [9]

To implement the proposed approach we have built a distributed architecture based on the following products: AMSEC, application server Apache Tomcat [2], DBMS Virtuoso [36], and security scanner MaxPatrol [26]. Elements of the techniques suggested have been implemented as services running on the application server in Java [31].

The AMSEC visualization subsystem provides visual tools for input data configuring and presenting results of attack modeling and security evaluation.

Let us represent the Network Constructor dashboard of the AMSEC used to setup initial data (Figure 5). It is divided into four subviews [11].

The main *view C* shows the topology of the studied network, while the *view A* reflects the hierarchical structure of the network, depicting domains or specified network zones. The graph based techniques are used to represent

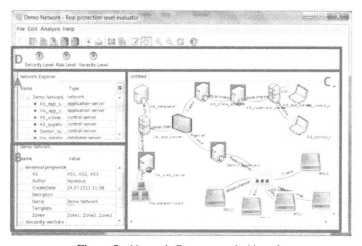

Figure 5 Network Constructor dashboard

network topology. Each network object is represented by an icon. The user has possibility to define icons for each type of the network objects. The background color of the icon is used to encode values of the security metrics calculated for the given host, such as Criticality, Mortality, Risk Level [13].

These metrics are chosen by the user from the predefined list. The brief information about each host is available via a tool tip which appears when mouse hovers over the network object.

The user can configure each host and network using the property *view B*. It can specify predefined properties of the host such as IP address, host type (web server, ftp server, database server, router, firewall, etc.), installed software and hardware, user-defined host criticality. These properties are necessary for attack graph generation. There is also a possibility to define user properties. This property view is updated whenever a particular state node is selected.

The *view D* shows the security metrics calculated for the network itself. As these metrics can have value from the predefined set of values {Low, Medium, Above Medium, High, Undefined}, they are presented in a form of the semaphore signal.

To depict the attack modeling results, we use graph based attack representation [11]. Each node of the graph denotes to specific attack action, and their order reflects the sequence of the malefactor actions: the nodes located on one level characterize actions that can be implemented simultaneously or independently from each other, while nodes located on different levels describe actions that are implemented in certain order.

The current version of the AMSEC prototype allows to use two levels of attack graph - network and host representation. At the first level, the graph is a visualization of a computer network with the transitions between hosts that indicate possible movements of malefactor. At the second level, the graph represents possible sequences of exploits (Figure 6).

This view could be useful when using color encoding of the security metrics of the attack actions, providing general impression on the attack complexity or severity. The tree view is more traditional and convenient when identifying the sequence of the malefactor actions.

6 Case Study and Experiments

We performed several experiments with the prototype implemented to show the advantages of the proposed framework.

The network for the case study "Managed Enterprise Service Infrastructures" [25] was selected.

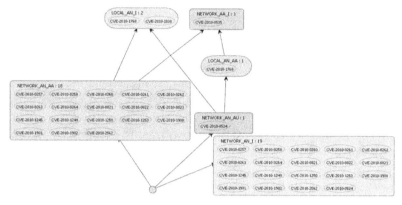

Figure 6   Attack graph for one host

The attack modeling and security evaluation process contained the following steps:

1. Preparation of data for constructing attack graphs (using source data collected from the network scanner);
2. Analysis of vulnerabilities inherent to hosts (taking into account software and hardware installed on each host);
3. Constructing attack graphs (based on data about attack actions available to the malefactor and about the network topology);
4. Analysis of attack graphs and security evaluation (for each host and for the whole network).

Then the changes in the network were simulated (about 10% of network objects were completely updated) and all steps were carried out again.

As a platform for experiments we used a computer with Windows 7 Service Pack 1 x64-based on quad-processor Intel i5 2, 3 GHz with 4 GB of RAM.

Figure 7 shows the dependency between the time required for different steps of attack modeling and security evaluation process and the amount of hosts in the network. The source network was generated randomly with condition that each host should contain at least one vulnerable software.

Experiments have shown that the vulnerability analysis (generation of possible attack actions) for hosts in the network is the most costly phase (in terms of execution time). Time spent on other phases does not exceed a few seconds.

The experimental results are averaged values. Analysis of these results gives an indication that the time required for construction and analysis of attack

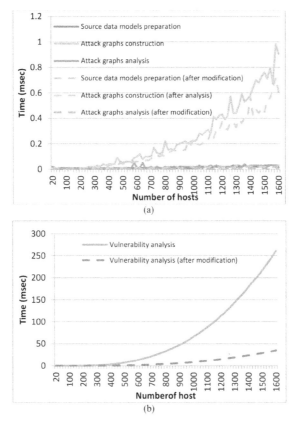

Figure 7   Approximate dependency between the time (sec) of AMSEC functioning and the number of hosts in the network

graphs for a computer network consisting of 1000 hosts does not exceed 2 minutes, and the time for complete renewal of 10% of hosts of the network (for this example it will be approximately 100 hosts) does not exceed 10 seconds. If the change affects only individual models (e.g. update of software on hosts), the necessary time can be greatly reduced. Thus, we can conclude that this approach can be used in systems that operate in near real time.

7 Conclusion

In the paper we presented the framework for computer attack modeling and security evaluation component (AMSEC). It outlines also the current prototype of the AMSEC on the whole and the implementation of the particular techniques for attack modeling and security analysis mechanisms. The AMSEC prototype was evaluated by several examples, and AMSEC successfully calculated the security metrics for them.

The results obtained make it possible to evaluate the security of computer networks in near real time, using techniques of analytical modeling of network attacks suggested in this paper. These results can be used to enhance the effectiveness of existing security mechanisms in SIEM systems.

All elements of attack modeling and security evaluation described in the paper will be considerably extended and detailed in the further research. Future research will be connected with enhancing of attack graphs building and analysis techniques. The analysis of the approach effectiveness assessment on real examples and extension of the set of security metrics will be performed.

Acknowledgment

This research is being supported by the grants of the Russian Foundation of Basic Research (13-01-00843, 13-07-13159, 14-07-00697, 14-07-00417), the Program of fundamental research of the Department for Nanotechnologies and Informational Technologies of the Russian Academy of Sciences (contract #2.2), and the project ENGENSEC of the TEMPUS program of the European Community.

References

[1] A. P. Moore, R. J. Ellison, and R. C. Linger, 'Attack Modeling for Information Security and Survivability', Technical Note CMU/SEI-2001-TN-001. Survivable Systems, 2001.

[2] Apache tomcat, http://tomcat.apache.org/

[3] B. A. Blakely, 'Cyberprints Identifying cyber attackers by feature analysis', Doctoral Dissertation: Iowa State University, 2012.

[4] B. Schneier, 'Attack Trees – Modeling Security Threats', Dr. Dobbs Journal, December, 1999.

[5] CAPEC. Common Attack Pattern Enumeration and Classification, http://capec.mitre.org/

[6] CAULDRON, http: //proinfomd.com/how.html.

[7] Common Platform Enumeration (CPE). http://cpe.mitre.org/

[8] Common Vulnerabilities and Exposures (CVE). http://cve.mitre.org/

[9] Common Vulnerability Scoring System (CVSS). http://www.first.org/cvss/

[10] E. Bursztein, 'Extending Anticipation Games with Location, Penalty and Timeline', LSV, ENS Cachan, CNRS, INRIA, France, 2008.

[11] E. Novikova, I. Kotenko, 'Analytical Visualization Techniques for Security Information and Event Management', Proceedings of the 21th Euromicro International Conference on Parallel, Distributed and network-based Processing (PDP 2013). Belfast, Northern Ireland. Los Alamitos, California. IEEE Computer Society, pp.519–525, 2013.

[12] I. Kotenko, A. Chechulin, 'A Cyber Attack Modeling and Impact Assessment Framework', Proceedings of the 5th International Conference on Cyber Conflict 2013 (CyCon 2013), IEEE and NATO COE Publications, Tallinn, Estonia, pp.119–142, 2013.

[13] I. Kotenko, A. Chechulin, 'Attack Modeling and Security Evaluation in SIEM Systems', International Transactions on Systems Science and Applications, Vol.8, December, pp.129–147, 2012.

[14] I. Kotenko, A. Chechulin, 'Common Framework for Attack Modeling and Security Evaluation in SIEM Systems', Proceedings of the 2012 IEEE International Conference on Green Computing and Communications, Conference on Internet of Things, and Conference on Cyber, Physical and Social Computing. Los Alamitos, California, IEEE Computer Society, pp.94–101, 2012.

[15] I. Kotenko, A. Chechulin, 'Computer Attack Modeling and Security Evaluation based on Attack Graphs', Proceedings of the IEEE 7th International Conference on "Intelligent Data Acquisition and Advanced Computing Systems: Technology and Applications" (IDAACS'2013), Berlin, Germany, pp.614–619, 2013.

[16] I. Kotenko, A. Chechulin, and E. Novikova, 'Attack Modelling and Security Evaluation for Security Information and Event Management', Proceedings of the International Conference on Security and Cryptography (SECRYPT 2012). Rome, Italy, pp.391–394, 2012.

[17] I. Kotenko, I. Saenko, O. Polubelova, A. Chechulin, 'Design and Implementation of a Hybrid Ontological-Relational Data Repository for SIEM systems', Future internet, vol.5, No.3, pp.355–375, 2013.

[18] K. Ingols, M. Chu, R. Lippmann, S. Webster, and S. Boyer. 'Modeling modern network attacks and countermeasures using attack graphs', Proceedings of the 2009 Annual Computer Security Applications Conference (ACSAC'09), Washington, D.C., USA, IEEE Computer Society, pp.117–126, 2009.

[19] K. J. S. Hoo, 'How much is enough? A risk-management approach to computer security', PhD thesis, Stanford University, CA, 2000.

[20] L. P. Swiler, C. Phillips, D. Ellis, and S. Chakerian, 'Computer-Attack Graph Generation Tool', Proceedings of the Second DARPA Information Survivability Conference & Exposition (DISCEX II), LosAlamitos, California, vol. II, pp. 307–321, 2001.

[21] L. Wang, A. Singhal, S. Jajodia, and S. Noel, 'K-zero day safety: measuring the security risk of networks against unknown attack', Proceedings of the 15th European conference on Research in computer security (ESORICS'10), Springer-Verlag Berlin, Heidelberg, 2010, pp.573–587.

[22] L. Williams, 'GARNET: A Graphical Attack Graph and Reachability Network Evaluation Tool', Proceedings of the 5th international workshop on Visualization for Computer Security, Springer-Verlag Berlin, 2008.

[23] M. M. Gamal, D. Hasan, A. F. Hegazy, 'A Security Analysis Framework Powered by an Expert System', International Journal of Computer Science and Security, vol.4, Issue 6, pp.505–526, 2011.

[24] M. McQueen, T. McQueen, W. Boyer, M. Chaffin, 'Empirical estimates and observations of 0-day vulnerabilities', Hawaii International Conference on System Sciences, 2009.

[25] MASSIF, 2013. Massif project, http://www.massif-project.eu

[26] MaxPatrol security scanner, http://ptsecurity.com/maxpatrol

[27] MITRE Corporation. http://mitre.org/

[28] N. Kheir, H. Debar, N. Cuppens-Boulahia, F. Cuppens, and J. Viinikka, 'Cost evaluation for intrusion response using dependency graphs', IFIP International Conference on Network and Service Security (N2S), IEEE, Paris, France, pp.1–6, 2009.

[29] National Vulnerability Database (NVD). http://nvd.nist.gov/

[30] Nessus scanner software. http://www.tenable.com/products/nessus

[31] Oracle Java SE? http://www.oracle.com/technetwork/java/javase/

[32] R. Dantu, P. Kolan, and J. Cangussu, 'Network risk management using attacker profiling', Security and Communication Networks, vol.2, No.1, pp.83–96, 2009.

[33] R. Lippmann, K. Ingols, 'Validating and Restoring Defense in Depth Using Attack Graphs', Proceedings of MILCOM 2006, Washington, DC, 2006.

[34] T. Olsson, 'Assessing security risk to a network using a statistical model of attacker community competence', Proceedings of the 11th international conference on Information and Communications Security, 2009, pp.308–324.

[35] The Center for Internet Security, The CIS Security Metrics, 2009.

[36] Virtuoso universal server, http://virtuoso.openlinksw.com/

[37] W. Kanoun, N. Cuppens-Boulahia, F. Cuppens, J. Araujo, 'Automated reaction based on risk analysis and attackers skills in intrusion detection systems', Proceedings of the third International Conference on Risks and Security of Internet and Systems (CRiSIS'08), Toezer, Tunisia, pp.117–124, 2008.

Biographies

Igor Kotenko is a professor of computer science and Head of Research Laboratory of Computer Security Problems of the St. Petersburg Institute for Informatics and Automation of the Russian Academy of Science. He graduated with honors from St.Petersburg Academy of Space Engineering and St.Petersburg Signal Academy, obtained the Ph.D. degree in 1990 and the National degree of Doctor of Engineering Science in 1999. He is the author of more than 200 refereed publications, including 12 textbooks and monographs. Igor Kotenko has a high experience in the research on computer network security and participated in several projects on developing new security technologies. For example, he was a project leader in the research projects from the US Air Force research department, via its EOARD (European Office of Aerospace Research and Development) branch, EU FP7 and FP6 Projects, HP, Intel, F-Secure, etc. The research results of Igor Kotenko were tested and implemented in more than fifty Russian research and development projects.

Andrey Chechulin received his B.S. and M.S. in Computer science and computer facilities from Saint-Petersburg State Polytechnical University, Saint-Petersburg, Russia, and PhD from St.Petersburg Institute for Informatics and Automation of the Russian Academy of Sciences (SPIIRAS). He is now a senior researcher at the Laboratory of Computer Security Problems of SPIIRAS. He is the author of more than 30 refereed publications. His primary research interests include computer network security, intrusion detection, analysis of the network traffic and analysis of vulnerabilities.

Code Search API, Base of Parallel Code Refactoring System For Safety Standards Compliance

Peter Jurnečka, Petr Hanáček and Matej Kačic

FIT BUT, Bozetechova 1/2 Brno, Czech Republic, {ijurnecka, hanacek, ikacic}@fit.vutbr.cz, www.fit.vutbr.cz

Received 3 February 2014; Accepted 27 April 2014;
Publication 2 June 2014

Abstract

New technologies of multi-core and massively parallel processors are becoming common parts of today's desktop computers. These state-of-the-art technologies allow programming of parallel applications and systems, however, creating parallel applications puts higher demands on programmers' skills, project maintenance and modification of existing source codes. Program flaws entered on source codes could have fatal consequences, specifically in aviation or medicine systems, due to possible fatal impacts in case of systems failure.

This paper describes the current status of aviation and medicine software safety standards, points out the common requirements of all these standards, specially the requirement for reliability. Reliability can be easily achieved using design patterns with verified reliable source code modules. In our research, we propose system for implementation of concurrency and synchronization design patterns into existing code. We have created parallel source code search API which is described in this paper, and which is planned to be used in our parallel code refactoring system for safety standards compliance. This API enables us to define appropriate places in source codes for introduction of parallel design patterns into existing parallel source codes. In next design iteration, the proposed system will provide suggestions of refactoring operations of found source codes, based on static code analysis and formal description of parallel design patterns.

Journal of Cyber Security, Vol. 3 No. 1 , 47–64.
doi: 10.13052/jcsm2245-1439.313

Keywords: software safety, parallel design patterns, code searching.

1 Introduction

Parallel or multithreaded applications are becoming more widespread. New technologies such as multicore processors and massively parallel processors of graphics cards have become widely available and usable in desktop computers. However, programming of parallel systems puts higher demands on the skills of programmers, and greater demands are also by the maintenance and modification of existing projects.

Area in which any mistake can have fatal consequences is aviation or medicine. Aviation safety standards [1, 2] play an important role, because failures may have fatal impact. When we speak about software in aviation, we mean software for avionics, which is a term used for electronic systems used in airborne environments, derived from words aviation and electronics. Examples of avionic systems used in aircrafts are flight control systems (autopilot), navigation systems or anticollision systems. Safety of the software is part of a whole system safety.

The purpose of the Food and Drug Administration (FDA) software validation standards [3, 4] is to consider its applicability to the validation of medical device software. The standards recommend an integration of software life cycle management and risk management activities. The software developer should determine the specific approach and level of effort to be applied based on these standards. On the other hand, FDA validation standards do not recommend any specific life cycle model or specific technique.

Avoiding mistakes is the goal of software standards in these areas. Other way to avoid mistakes is to facilitate the work of programmers by using design patterns and refactoring. Currently, much research has been done in the field of design patterns and refactoring of existing source code. However, the research of automated refactoring has not addressed design patterns of parallel and distributed systems.

Common requirement of all these standards is the requirement for reliability which can be achieved with design patterns. In our research, we propose system for implementation of parallel design patterns in existing code. The proposed system provides suggestions of refactoring operations based on static code analysis with code search API and formal description of parallel design patterns.

The main idea of proposed system is to use formally specified parallel design patterns in suggesting refactoring operations in editing of source code. The aim of our research is to create a system that automatically assist the programmer in source code refactoring in implementing parallel design patterns into existing parallel source code. Source code created with use of design patterns is more efficient, easily manageable and therefore more reliable. To create such system we must combine parallel design patterns, refactoring and static code analysis.

2 Code Searching Problem

The issue of searching source codes has been given a great amount of research. There are different approaches used, each approach has its advantages, but in our context of the definition of insertion places of design patterns neither cannot be used, because none of these existing solutions do not search parallel source code and therefore has no information about access of program threads to each source code statement. This is our main contribution delivered by this article.

First existing code search solution is XL C++ Browser from [5], which is a distributed static analyzer for the C++ programming language. Key features of this technology are its support for semantic queries - queries that make use of the C++ semantics to interpret information about programs. It uses rules for describing the relations between the program symbols and it has capability to browse remote databases across network.

Steven P. Reis in his Semantics Based Code Search [6] describes his system which uses the vast repositories of available open source code to generate specific functions or classes that meet user specifications. He lets users specify what they are looking for as precisely as possible using keywords, class or method signatures, test cases, contracts, and security constraints. His Code Search system then uses an open set of program transformations to map retrieved code into what the user asked for.

This approach was implemented in prototype system for Java with web interface. Limitation of this solution is, that it is very tightly bound to java and generalization of search engine to other languages and implementation of thread info is more difficult than creating a new API.

Lemos in his article Applying TestDriven Code Search to the Reuse of Auxiliary Functionality [7] states that software developers spend considerable effort implementing auxiliary functionality used by the main features of system (e.g. compressing/decompressing files, encryption/description of data, scaling/rotating images). With the increasing amount of open source code available on the Internet, time and effort can be saved by reusing these utilities through informal practices of code search and reuse. However, when this type of reuse is performed in an ad hoc manner, it can be tedious and errorprone: code results have to be manually inspected and extracted into the workspace. In his paper he introduces the use of test cases as an interface for automating code search and reuse and evaluate its applicability and performance in the reuse of auxiliary functionality. He calls his approach TestDriven Code Search (TDCS). Test cases serve two purposes: (1) they define the behavior of the desired functionality to be searched, and (2) they test the matching results for suitability in the local context. He presents CodeGenie, an Eclipse plugin that performs TDCS.

CodeGenie is most similar to the proposed solution, however as discussed above, CodeGenie does not search parallel source code and does not include information about threads and their access to source code statements.

Last found solution is Sourcerer [8]: Search Engine for Open Source Code Supporting Structure Based Search: The paper [8] focuses on the current research goals and search capabilities of Sourcerer. Sourcerer enables searches that are based not just on keywords but also on the structural properties and relations among program elements. Current version of Sourcerer works with the open source projects implemented in Java. The first release of Sourcerer, as of time of submission of that paper, is publicly available in its development version at http://sourcerer.ics.uci.edu/.

Sourcerer is closest to our approach. But because it uses large relational database as data storage and is closed only to Java and also has no information about threads, it is not usable in our context. Therefore we propose own code search API and system, which will be used to define the search queries for parallel source code. This queries will find suitable places in the code which we can propose to add a design pattern.

3 Our Solution

Our Code Search API provides interface for easily searching for specific code. It is one of base parts of our parallel code generating and refactoring system for safety standards compliance, which consists of two parts, design pattern code generator and the new code structure proposer. The code generator generates source code from existing source code using formally described design patterns. The code structure proposer founds appropriate places in code for applying design patterns using Code Search API and selected design patterns are then used to generate new source code containing code snippets from the original code.

As mentioned before, Code Search API provides interface, which is used for queuing existing source codes. This interface is used in our proposed design pattern specification language. Each design pattern is described with pair (*prec, spec*) where prec stands for precondition and spec for pattern specification.

Precondition defines places, where concrete design patterns should be placed. This is done using our proposed Code Search API, which provides robust language for static parallel code analysis used for determining appropriate design pattern usage. Whole static code analysis is running on our code searching framework which provides easy access to all classes, functions, properties in project including thread usage information for each of theese statements.

Pattern specification (spec) uses XML which consists of two main parts: entities (denoted by <entities>) and relations between them (denoted by <relations>). Part <entities> contains a definition of entities that exist in design patterns. Each entity can represent class, method, variable or property. Entities may contain attributes which determine the connection of the design pattern specification *spec* with each statement from the set of preconditions *prec*.

Main idea behind our Code Search API is extension of abstract syntax tree with thread usage information. This thread info contains information about threads for each line of code in source files. Source code is parsed into syntax tree and during this iteration, *ThreadStartFilter* filter is applied on each statement. If selected statement creates new thread, then new thread is added into Thread List. We have implemented C# prototype, and for example, our ThreadStartFilter returns all known thread starting statements (e.g. *Thread.Start(), Task.Factory. CreateAndStart(), Background Worker*).

4 Metamodel used for Modelling AST

There are multiple approaches for storing source code in a queryable repository, some are based on abstract syntax trees (ASTs) and others on relational databases. We did not want to create an entirely new metamodel for Code SearchAPI, but rather extend an existing one that met our requirements. It had to be sufficiently expressive as to allow structure-based analyses, and it had to be efficient and scalable enough to include thousands lines of source code. We settled on an adapted version of Ossher et al.'s [10] SourcererDB which was based on Chen et al.'s [11] C++ entity-relationship (ER) metamodel. While in principle as expressive as an DB-based metamodel, our object metamodel is better suited to define search queryes which are a cornerstone of CodeSearchAPI. In addition, we agreed with their decision to focus on what they termed a top-level declaration granularity, as it provides a good compromise between the excessive model size of finer granularities and the analysis limitations of coarser ones. The metamodel we present here is an extended and modified version of Sourcerer metamodel [10]. As shown on Figure 1. our revised metamodel adds support for thread information and removes Java bindings, which are not usable in our environment. This metamodel is used to model the structure and reference information extracted from .Net C# projects. Each source code file contains the entities defined within it, the relations originating from those entities, and the comments associated with them.

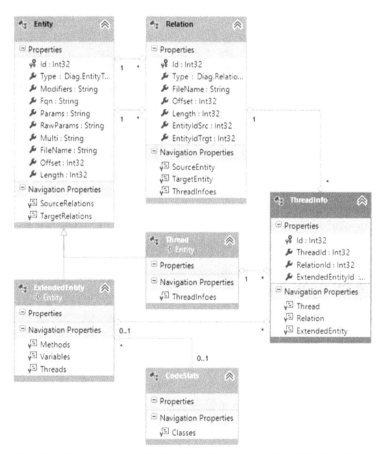

Figure 1 Extended syntax tree with thread and type info, collected by Code Search API

Base of our system is taken part of Sourcerer DB metamodel: table entites and relations. Our classes based on theese tables are created using Sourcerers proposed algorithms. On this basis, we have build our extended data structure, which contains classes: Extended Entity, Thread, Thread Info. Code Stats class serves as common unique entry point for all queries, and enables simple unified approach to search queries creation. The folowing paragraphs describe each class of our system, their creation and their purpose in our system.

4.1 Entity Class

Class Entity is taken from the design of Sourcerer. However, in our solution, we have removed specific java entity types ajd java bindings unusable in our system and we also have simplified the system for linking entities to files. The majority of entity types used in our metamodel correspond to explicit declarations in the .Net C# source code and shoul also be enough for all major object oriented languages. The entity types are: NAMESPACE, CLASS, INTERFACE, ANNOTATION, FIELD, INITIALIZER, CONSTRUCTOR , METHOD, PARAMETER, LOCAL VARIABLE, ARRAY, TYPE, PRIMI-TIVE, ENUM, ENUM CONSTANT, INSTRUCTION. Each entity is uniquely identified by its Fully Qualified Name (Fqn : String), file that it comes from, and its location in that file. Each entity is further annotated, when appropriate, with its modifiers (such as public or static).

4.2 Relation Class

Class Relation is also taken from the design of Sourcerer, however in our application we have removed relation types, which are not usable in our search queries. Table 1 contains the relation types in our metamodel. All of the relations are binary, linking a source entity with a target entity. A relation is identified uniquely by its type, and the FQNs of its source and target entities and source code location. As a result, any time the same relation is generated more than once, such as a method calling another method multiple times in loop in its body, those relations are collapsed into one.

4.3 Thread Class

Thread class is a special kind of Entity used to tracking of the threads in the application. Each thread is basically a special kind of method, which is called parallel with the main method of the program. As mentioned above, during the creation of AST the ThreadStartFilter is applied to each command and

Table 1 Relation Types

Relation	Description
CONTAINS	Physical containment
IMPLEMENTS	Interface implementation / extension
TYPE OF	Field type
RETURNS	Method return type
READS	Field access
WRITES	Field access
CALLS	Method invocation
INSTANTIATES	Constructor invocation
THROWS	Throws declaration or explicit throw
ANNOTATED BY	Annotation
USES	Any reference
PARAMETRIZED BY	Associated type variables
OVERRIDES	Function overrides

if it detects a new thread this thread is tracked using this special entity type Thread. For our purpose (search queries for automated insertion of paralle design patterns) we need to divide Entites into 3 groups: those never touched with any thread, those used with one thread and those used with two or more threads. This division simplifies the process of Thread detection. If some new thread is created in loop, or in recursive function call, we automaticly create two instances of new thread, because that is the worst case. As mentioned before, Thread as subclass of Entity is uniquely identified by its Fully Qualified Name and location in source code. When we during AST traversal come to already tracked location on source code, we create new Thread and continoue, only when there is only one Thread object created, Otherwise we skip this code as already covered. In this state of our research, we do not take into account thread differences, caused by parameters used by thread creation code.

4.4 ThreadInfo Class

ThreadInfo class is used for the definition of relation between Threads and Entities. They tell us that selected thread touches variable in this function call. ThreadInfo class are markes all Relatons associated with each line of thread source code line. ThreadInfo classes are created during the construction of Thread entities, by taking all Relations related to main thread function. If the main thread function calls other functions, also those relations are marked.

4.5 ExtendedEntity and CodeStats Classes

ExtendedEntity and CodeStats classes simplify the definition of search queries. CodeStats class is a single common entry point for all queries. ExtendedEntity classes create tree structure and are formed during the creation of Entities during first pass of code parsing and creation of AST. If the type of Entity is class or method or property the ExtendedEntity is created. ExtendedEntity class extends Entity with three additional navigation properties which are Methods, Variables and Threads. Property Methods is used only when ExtendedEntity points to class and contains references to all methods and constructors within class. Property Variables contains information about all shared variables within selected ExtendedEntity. Property Threads links ExtendedEntity with ThreadInfo and is its main purpose is to make querying easier. As we shoved in next chapter, this metamodel is sufficient for our proposed specification of design patterns and their automatic insertion into existing source codes.

5 Results

We have applied this system to set of six synchronization patterns from the POSA catalogue [9]. In almost all cases it is possible to find reasonable precondition. Next paragraphs provide short description, UML diagram and found preconditions of selected patterns described by Microsoft LINQ queries into our CodeSearch API data model. Possibility of definition of query using our CodeSearch API showes, that CodeSearch API can be used in next iteration of our code refactoring research. To give ourselves coarse outline, in next subchapters we also provide easily readable version of preconditions of selected synchronization design patterns. All these patterns are used for avoiding synchronization problems between threads.

5.1 Thread Safe Interface

As shown on Figure 2 Thread - Safe Interface pattern divides the functions of the component to the public available interface and private implementation methods. Public interface acquires the lock, calls corresponding private method and then releases the lock.

To ensure proper synchronization of client threads call functions of only the public interface. Interface function obtains necessary locks and calls the appropriate implementation of the function that no longer cares about locking and can freely call other implementation functions. Deadlock (selfdeadlock) cannot occur because the lock is obtained only once at the beginning in the

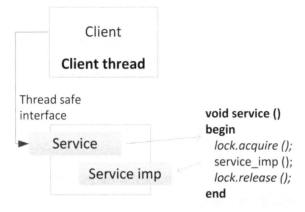

Figure 2   Thread - Safe Interface design pattern

```
CodeStats.Classes
.SelectMany(type => type.Functions)
.Where(func => ((func.Threads.Count > 1)
&& (GetLocks(func).Count() == 0));
```

Figure 3   Thread - Safe Interface precondition code

public interface and the recursive function calls do not obtain the lock more times, which also improves the performance of the application.

From this description we can define the preconditions of this pattern, which are: There exists an object with one or more functions or properties, which are used by more than one thread and this function or property does not acquire any lock. As we can see on next code snippet, this precondition uses ThreadInfo, from CodeSearch API, and function GetLocks(ClassMember x) which gets locks used by selected member of class. On Figure 3. we can see Microsoft LINQ source code version of precondition of this pattern.

5.2 Future

As shown on Figure 4 the Future pattern immediately after calling the constructor returns the "virtual" data object, called the Future. Future object contains information about the state and the calculation of the results is dispatched on service thread. The future object returns the result only if it is valid.

If the client thread wants to read the future value of the object before there is a valid result, the future object suspends client thread until a valid result

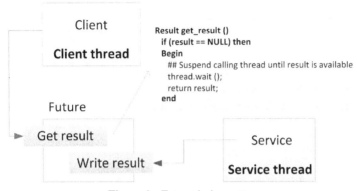

Figure 4 Future design pattern

```
CodeStats.Classes
.SelectMany(type => type.Functions)
.Where(func =>
AvgFunctionLen(func) >= treshold);
```

Figure 5 Future design pattern precondition code

is written in the future object. Future object can also contain nonblocking function available for checking the validity of stored value.

From this description we can define the preconditions of this pattern, which are: There exists object X, this object contains function or parameter with execution time longer than user defined threshold. Or, there exists object X with constructor with execution time longer than user defined threshold, and first call of parameter or function of this object is far away from its construction. As we can see on next code snippet, this precondition uses function AvgFunctionLen(ClassMember x) which gets average function length of selected member of class, which is computed by counting executed instructions. For safety and calculability reasons AvgFunctionLen counts only to UInt16.MaxValue. On Figure 5. we can see Microsoft LINQ source code version of precondition of this pattern.

5.3 Guarded Suspension

As shown on Figure 6 Guarded Suspension design pattern instead of termination of the blocked functions suspends the thread, so that other threads can access shared component and thus change the value of the guard conditions and the release threads of the blocked functions.

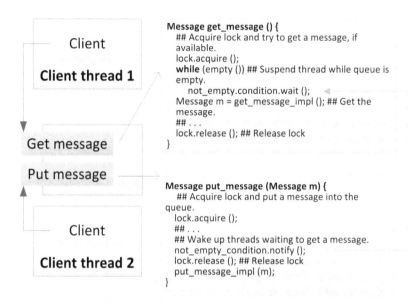

Figure 6   Guarded Suspension design pattern

The main contribution of Guarded Suspension design pattern is that it minimizes the costs associated with parallelization and also increases the availability of shared components. If threading model is designed to make suspension on OS layer, the price of the suspension and the synchronization is minimal.

From this description we can define the preconditions of this pattern, which are: There exists function that is dispatched in separate thread, and this function contains condition which aborts functions thread without any other computation. As we can see on next code snippet, this precondition uses function InstrCounter(ClassMember x) which creates pairs instructionId, instruction. For safety and calculability reasons InstrCounter counts only to UInt16.MaxValue. On Figure 7 we can see Microsoft LINQ source code version of precondition of this pattern.

```
CodeStats.Classes
.SelectMany(type => type.Functions)
.Select(func => InstrCounter(func))
.Where(instr =>
(instr.Id <= treshold) && (instr.IsReturn));
```

Figure 7   Guarded Suspension design pattern precondition code

Figure 8 Scoped Locking design pattern

```
CodeStats.Classes
.SelectMany(type => type.Variables)
.Where(v => v.Threads.Count() > 1)
.Where(v => GetLocks(v).Count() == 0);
```

Figure 9 Scoped Locking design pattern precondition code

5.4 Scoped Locking

As shown on Figure 8 the Scoped Locking design pattern outlines the critical section with lock statement which automatically gets a lock at the entrance, and automatically releases the lock on any way from the lock frame. Scoped Locking design pattern increases the robustness of parallel software by eliminating common programming errors associated with synchronization of multiple threads. Locks are obtained automatically when the thread enters the critical section, and automatically released when it leaves out. Implementation of this design pattern depends on the programming language. For example, the Java programming language contains the synchronized keyword which instructs the compiler to automatically generate the appropriate instructions serving locking and unlocking locks.

From this description we can define the preconditions of this pattern, which are: Number of shared variables among the program is small, and access to these variables is not secured with any locks. As we can see on next code snippet, this precondition uses ThreadInfo, from CodeSearch API, and function GetLocks(ClassMember x) which gets locks used by selected member of class. On Figure 9 we can see Microsoft LINQ source code version of precondition of this pattern.

5.5 Immutable Value

As shown on Figure 10. Immutable Value design pattern defines the design objects whose instances are immutable. The internal state of an object is set in the constructor, and no further changes are allowed.

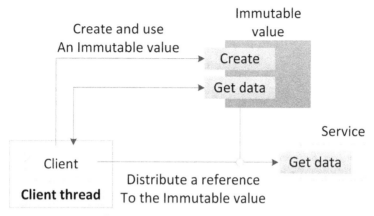

Figure 10   Immutable Value design pattern

```
CodeStats.Classes
.SelectMany(type => type.Variables)
.Where(v => GetWritesCount(v) == 1);
```

Figure 11   Immutable Value design pattern precondition code

In an immutable object are only read only parameters. The absence of any possibility of changing the object removes any need for synchronization and thus simplifies and improves efficiency the work of the program. By eliminating the need to copy objects we also improve program performance.

From this description we can define the preconditions of this pattern, which are: All variables, that are only read in program execution, or are written only one time. As we can see on next code snippet, this precondition uses function GetWritesCount(Variable x) which gets number of writes into this variable during program execution. On Figure 11 we can see Microsoft LINQ source code version of precondition of this pattern.

6 Conclusions

The importance of safety standards of software systems is increasing as the use of software grows because of its convenience and flexibility. Software safety standards are very important in aircraft, military, automotive or medical devices. Common requirement of all standards is reliability. Reliability can be easily achieved with design patterns. We are creating system for implementation of parallel design patterns into existing code. Our system will provide suggestions of refactoring operations based on static code analysis and formal description of parallel design patterns. For this purpose, we have created custom CodeSearch API, which queries will be used in design patterns definitions, for defining appropriate places for design patterns suggestions. The last part of article gives us of an overview of application of created CodeSearch API on a set of six synchronization patterns from the POSA catalogue [9]. In all cases it was possible to find reasonable precondition using our Search API. In next iteration we focus our research on better customizable XML description language, an finalization of whole refactoring system, so we can then provide full robust parallel design patterns refactoring system for safety standards compliance.

This work has been supported by the European Regional Development Fund in the IT4Innovations Centre of Excellence project (CZ.1.05/1.1.00/02.0070) and by BUT FIT grant FIT-S-11-1: "Advanced secured, reliable and adaptive IT" and by Research Plan No. MSM0021630528.

References

[1] Howard C. (2011). DO-178B safety certification and other software security tools drive avionics software designs. 2011. Available at: http://goo.gl/ZkzyF

[2] Federal Aviation Administration. Advisory Circular 20-115B. 1993. Available at: http://goo.gl/C6d1k

[3] General Principles of Software Validation; Final Guidance for Industry and FDA Staff, Available at: http://goo.gl/HjIKb

[4] Guidance for the Content of Premarket Submissions for Software Contained in Medical Devices, Available at: http://goo.gl/JqkYr

[5] SHARMAN, J. ET AL. (1992) Architecture of the XL C++ browser, CASCON '92 Proceedings, P: 369–379, IBM Press

[6] REISS, P. STEVEN, (2009) Semantics-Based Code Search, ICSE 09 Proceedings, IEEE

[7] LEMOS, OTAVIO AUGUSTO LAYYARINI, ET AL. (2009) Applying Test-Driven Code Search to the Reuse of Auxiliary Functionality, SAC 09 Proceedings, ACM

[8] SUSHIL B., ET AL. (2006) Sourcerer: A Search Engine for Open Source Code Supporting Structure-Based Search, OOPSLA 06 Proceedings, ACM

[9] BUSCHMANN, F. ET AL. (2007) Pattern-Oriented Software Architecture: A Pattern Language for Distributed Computing. John Wiley & Sons, Inc., New York, NY USA, ISBN: 978-0-470-05902-9.

[10] J. Ossher, S. Bajracharya, E. Linstead, P. Baldi, and C. Lopes, "SourcererDB: An aggregated repository of statically analyzed and cross-linked open source Java projects," in Proceedings of the International Workshop on Mining Software Repositories. Vancouver, Canada: IEEE Computer Society, 2009, pp. 183–186.

[11] Y.-F. Chen, E. R. Gansner, and E. Koutsofios, "A c++ data model supporting reachability analysis and dead code detection," IEEE Trans. Softw. Eng., vol. 24, no. 9, pp. 682–694, 1998

Biograpies

Peter Jurnečka. He received his M.Sc. from Brno University of Technology in 2009. He is currently a Ph.D. student at Faculty of Information Technology, Brno University of Technology. His research interests are in information technology security and safety, especially in using parallel design patterns for safety standards compliance.

Petr Hanáček. He graduated at Brno University of Technology. He is currently an Associate Professor at Faculty of Information Technology, Brno University of Technology. His research interests are in security of information systems, applied cryptography and wireless systems.

Matej Kačic. He received his M.Sc. from Brno University of Technology in 2010. He is currently a Ph.D. student at Faculty of Information Technology, Brno University of Technology. His research interests are in information technology security, especially in wireless systems.

Memory Acquisition by Using Network Card

Štefan Balogh

Štefan Balogh, Slovak University of Technology, Faculty of Electrical Engineering and Information Technology, Ilkovičova 3, Bratislava SK-812 19, Slovak Republic, stefan.balogh@stuba.sk

Received 4 March 2014;; Accepted 27 April 2014;
Publication 2 June 2014

Abstract

To detect present rootkit the rootkit and malware detectors need to have memory access. But, sophisticated rootkits are able to subvert the verification process of security scanner using virtual memory subversion techniques to hide their activity. We have proposed a new solution for direct memory access, based on a custom NDIS protocol driver that can send (on request of the local executable program) the contents of the computer memory over the network. Our method allows an unexpected type of the direct memory access, which is independent of the processor, and its control capabilities. This is a strong advantage in rootkit detection, because the rootkit cannot take any action to hide itself while the memory is scanned.

Keywords: Live Forensics; Memory Acquisition; DMA; Forensic analysis; network card; direct memory access; rootkit detection.

1 Introduction

Rootkits have become very sophisticated over the past few years. In 2006 the prototypes of rootkits that can subvert even the operating system by targeting hardware and firmware were presented in [1]. We have seen a surge in rootkit deployments in spyware, worms, and botnets. Generally, there are two types of rootkits: persistent rootkits and memory-based rootkits. Unlike persistent rootkits which can be loaded from disk into memory after reboot, in-memory rootkits make no effort to permanently store their code on disk or hook into

Journal of Cyber Security, Vol. 3 No. 1 , 65–76.
doi: 10.13052/jcsm2245-1439.314

the boot sequence. Their code exists only in volatile memory and they may be installed covertly via a software exploit. Thus kernel rootkits can control the execution path of kernel code, alter kernel data, and fake system call return values. Although once a computer system has been subverted by a rootkit it is extremely difficult to detect or eradicate the rootkit, there are still some different methodologies that detect the rootkit that have worked to varying degrees. To detect the kernel rootkits, we can use different techniques.

Scanning a signature of the rootkit during its in-memory execution, are worth mentioning because they have been applied with success to scanning system memory in addition to file system scanning. Ironically, most public kernel rootkits are susceptible to signature scans of kernel memory. It also can solve the problem with camouflaged binaries (by using a packing routine). But, these detection technique are useless against malware and rootkits for which a known signature does not exist. More-sophisticated rootkits are able to subvert the verification process by presenting an unmodified copy of the file for inspection, or by making modifications only in memory. So, memory scan based detection methods are useless against Virtual Memory Manager (VMM) hooking rootkits like Shadow Walker which are capable of controlling the memory reads of a scanner application [2]. Shadow Walker, can fake the contents of memory seen by other running applications. When the detector attempts to read any region of memory modified by the rootkit, it sees a 'normal', unaltered view of memory. To detect this kind of rootkits the memory scanner cannot depend on the OS API function or on processor control mechanisms.

Forensic analysis fights with similar limitations as rootkit detection: live detection can almost always be defeated by resident rootkits [3]. On the other hand offline analysis of the memory allows an investigator to see the state of the operating system without the operating system as a filter. Unlike trying to find a rootkit on a live system, the rootkit is unable to take any action to hide itself in the memory image. Therefore, investigators can see the data without the operating system, or the rootkit, interpreting the data for them [4].

There are also other methods today for rootkit detection. The earlier mentioned Signature based detection, or Heuristic / Behavioral detection (used by tool like VICE or Patchfinder [5]). Relatively new is cross view based detection which show a lot of promise. But, its success depends in large part upon implementation, specifically the method which is used to obtain the "low level" view of the system. Integrity based detection provides an alternative to both signatures and heuristics. However, the integrity checker is usually not capable of pinpointing the origin of the activity that has caused the changes.

Kernel rootkits can be especially difficult to detect and remove, because they operate at the same security level as the operating system itself and thus are able to intercept or subvert the most trusted operating system operations. To detect this generation of rootkits we need access to memory. The need for memory scanners are not confined to rootkits, but can also help to fight techniques that viruses started use since 2001, e.g. worm W32/CodeRed or W32/Slammer. These worms don't attack the files and computer disk. The malicious code can be executed before it is saved to disk. The worm is active only in memory and does not create any file. For this reason, neither files scanners nor on-access scanners (with reactive approach) are able to detect these attacks [6].

All existing rootkit detection techniques, it becomes apparent have strengths and weaknesses. The first step for rootkit and memory malware detection is scan the physical memory of the computer looking for rootkit behavior such as hooks in the SSDT, alterations to kernel functions (using kernel integrity checks), and modifications to key data structures like those performed by a DKOM attack. The tool for memory scan must remain as independent of the potentially subverted operating system as possible. To achieve this we:

- need access to memory (into different regions) without OS intervention
- must be able to make copy of the whole memory (memory dump – forensic analysis)
- and it's better to store the memory dump or part of memory copy in another computer)

In this article we have proposed a new method for scan the contents of the memory. The contents of memory are read by using DMA, so we avoid the possibility to change the contents by the running code. It could be very helpful for forensic purpose and also for rootkit detection.

The article is divided as follows: in the next section we present an overview of the methods for memory acquisition. Then we describe the existing attack that accesses the memory using a network card. In the third section we describe our new approach. Finally, we discuss the limitations and possible future improvements of the method.

2 Memory Access

Typically, there are two main approaches to memory access: software and hardware oriented, respectively. Software oriented tools use the fact that many operating systems allow reading the contents of memory: virtual devices dev/mem and dev/kmem in UNIX, and DevicePhysicalMemory in Windows. However, even the act of running a memory dumping tool itself changes a portion of RAM. When the program is loaded into memory, it can (due to memory paging) move useful information to the page file [7]. But, the result, seen by the tools, can by modified by resident rootkits. Due to these shortcomings, software solutions for memory acquisition are not reliable.

The main idea of hardware oriented approaches is to bypass the operating system using a physical device. The device creates a memory access through a single channel, which is independent of the operating system. Very often firewire interface [8] or DMA (Direct Memory Access) feature is used for direct access. This allows us to obtain a memory image without launching another process or having to use potentially contested features of the local system. A concept of special-purpose PCI device can be used either for forensic purpose [9] or for rootkit detection [10], [11]. But, these devices either must, be already present in the system at a time, when we need to access the memory contents [9], or need to support stand-alone mode [10] (which may not be supported by many commercial network cards).

Several attacks to gain control of the DMA, and read the contents of RAM were presented. Almost exclusively these attacks are based on network cards. A network card uses DMA, and also allows a network access, so the content can be sent directly to the recipient. The first attack of this type was presented by a team from the French Office for Information Security (Agence Nationale de la Sécurité des Systèmes d'Information, ANSSI) in 2010 [12]. They abused the vulnerability of network cards from Broadcom, which had an incorrect implementation of ASF (Alert Standard Format) version 2.0, and exploited a bug in RISC RX (receiving parts of the network card).

Another way to obtain an image memory by the NIC was presented in 2011 by Guillaume Delugré [13]. It uses the fact that if we are able to run modified firmware on computers with network cards from Broadcom, we are able to control the packets, and thereby control what is read and written using DMA. The attack itself starts by changing the BD used by DMA write. Delugré found how DMA write can change the pointer in the memory, size of the data and flags. Necessary condition for this attack is to have a modified firmware for the adapter, which is not publicly accessible. In order to be able to send data remotely we need also, a special firmware with ASF support.

In 2011 Jiang Wang et.al [14] proposed firmware assisted method to reliably acquire the memory. They combined a commercial PCI network card (which is widely available) with the System Management Mode (SMM) to acquire the memory and CPU registers. They also change the SMM code, and to solve the problem with compromising the network cards drivers by malware, they put the drivers into the SMM. But, in real we can write our own code and load it into the SMM only for some old machines, where the SMM is not locked by the BIOS. For new machines, the SMM is typically locked by the BIOS [14].

3 Memory Copy Using NDIS Driver

Our ultimate goal is to have a generic tool that can be used to read the memory contents for detection of rootkits, or for forensic purposes. None of the previous approaches is currently quite suitable due to its complexity, or a limited availability (FireWire). Therefore, we decided to create our own system. Our inspiration comes from Delugré attack, but we have chosen a different way to initialize DMA transfer, concretely the superstructure of network adapters NDIS (Network Driver Interface Specification). NDIS protocol driver can be used to service any network card.

The main goal of the memory access code is to create the NDIS driver packet with header and data section. The header contains the necessary parameters according to the used communication protocol (IP, TCP, UDP) and the data section contains the requested part of the memory. The packet is sent to the network driver and the network driver by using a standard communication

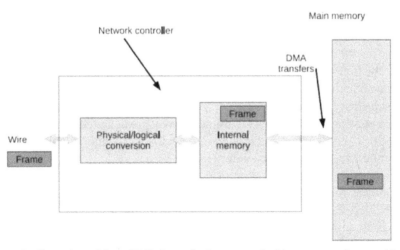

Figure 1 Ilustration of how DMA is used when network driver communicates with the network card (NIC) [13]

channel through DMA (see Figure 1) sends this packet to NIC. This DMA transfer avoids the possibility to change the contents of the memory by OS API functions or by processor control mechanisms when OS is compromised by malware.

We can use a remote control program or a local control program installed on the target computer to set up the necessary parameters for packet sending (IP address or MAC address where the memory content will be sent, and a required memory area, respectively). The remote control program can communicate with NDIS driver through the network. We have created a simple custom protocol for the communication between the control program and the driver.

3.1 NDIS Driver

NDIS driver is an interface with the main purpose of providing a standard API for the network interface card (NIC). It acts as an interface between the second and the third layer. The basic types are a protocol driver, an intermediate driver, and a miniport driver, respectively. Each has a different function and operates at a different level of the packet sending process. Network traffic, which is accepted by the NIC is controlled by the miniport driver. The protocol driver implements various protocols such as TCP/IP protocol. Finally, the intermediate driver is a hybrid between the previous two mentioned types. It is located between the Data link and Network layer and can control the operations accepted by NIC.

The intermediate and miniport driver has the possibility to control all incoming packets. So if a special packet is created and sent to the controlled computer, the NDIS driver will detect this packet, and setup the parameters, respectively. In this way we can remotely control the computer. For our practical implementation we have chosen the protocol driver, which is sufficient to provide a desired type of memory access. Also, remote control is not required in this test version. NDIS packets within the driver are declared as a structure of type NDIS_PACKET. We do not go into details, as the individual parts of that structure would not be treated directly.

The basic principle of operation of the system in the NDIS packet could be described as follows:

The NDIS_PACKET structure serves as a guiding structure which contains the basic information about the composition of the data. It is connected to the data in a buffer declared as NDIS_BUFFER (real data from higher layers to be sent), see Figure 2. Each NDIS_PACKET can be connected to several NDIS_BUFFER's.

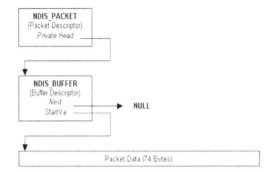

Figure 2   A simple version of NDIS buffer chaining [15]

3.2 Getting Memory Content

NDIS_PACKET structure contains information about connected buffers. Address of the sending packet in memory is located in NDIS_BUFFER structure. By using Ethernet II frame and IP protocol our packet has its own header and data section. Our goal is to replace the data section with a specified memory area, but preserve the Ethernet header (otherwise the packet will be lost in the network).

We divide a buffer to 2 smaller ones, and so we are able to work separately with the data, and with their memory space, respectively. The basic principle of the proposed approach is in Figure 3.

The attack proceeds as follows:

1. A new buffer pNdisBuffer1 is created. It has 14 bytes. In other words, we create a buffer in which the Ethernet header is stored.
2. When the data are separated from the header, we can modify them in such a way that they point instead of the Ethernet packet data to a designated address in the memory.
3. The buffer variable (pointer) that stores the address of data that contains the buffer is called MappedSystemVa. When we change its value to some arbitrary location in memory, the system will think that it contains the actually allocated data.
4. Setting the variable MappedSystemVa from buffer pNdisBuffer2 to the address of the beginning of memory space, or another requested memory location.
5. We are preparad for sending a packet with a selected chunk of memory. It remains only to combine the parts into one unit, which is done by concatenating pNdisPacket with pNdisBuffer1, and pNdisBuffer2.

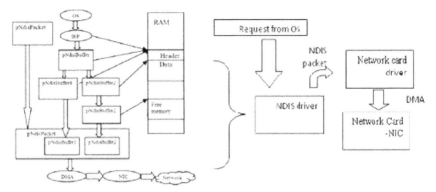

Figure 3   Basic principle of reading and send the memory contents

6. Finally, it remains only to send the packet. This will be served by the network driver and DMA engine.

But, in this way we can acquire only the access the memory contents or get the whole physical memory image of the target machine. To analyze or protect the critical OS components and structures we need to know their physical address. Also, to understand the physical memory, we must translate it to virtual memory used by the OS and find out the semantics of the memory. Only with this knowledge we can more effective protect or monitor the OS. However, it is difficult to obtain the semantics of a memory dump without knowing the values of the CPU registers at the time that the dump was retrieved. For that purpose, two CPU registers are critical: IDTR and CR3 [14]. IDTR points to the current interrupt descriptor table (IDT) and CR3 points to the base address of the current page table. CPU registers can be obtained in our solution using software based method. For this purpose we can utilize our NDIS driver. We can create special functions for getting information about CPU registers and also for getting the physical address of critical OS components and structures. This information's can by send as special packet to remote control program or a local control program. To implement extension like this make this solution more useful. We are actually working on this kind of extension.

3.3 Testing

To verify that we can really read the contents of memory from the desired memory address, we can use a simple technique in which known data is stored in memory. If you read them correctly back, it means that you really

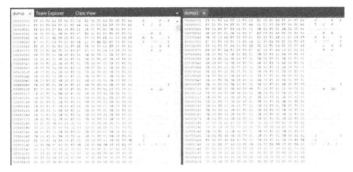

Figure 4 Memory dump: comparison of our solution and MDD tool

read your computer's memory. We try this test with data stored in known memory address with success. We have also tried to compare our dump with the results obtained by other existing tools. We made comparison, with dump of memory obtained by "MDD tool" [16]. MDD is capable of copying the complete contents of memory on computers with Microsoft Operating Systems. Both dumps were exactly the same (see Figure 4). These two kinds of tests indicate the possibility to get whole memory or only the specific part of the memory using our approach. However, this testing was performed on a clean PC without the rootkit. Testing with rootkits is more complex and must have correct methodology. We plan to test our method with the selected well-known rootkits of different types.

4 Conclusion

This paper deals with the study and implementation of direct memory access that satisfies the requirements for rootkit detection and for forensics analysis. In these cases we need to get the data from memory without a possible intervention of the operating system, or some rootkit installed on the computer. Therefore it is appropriate to implement a memory access in a way which is not controlled by the processor. Thus the rootkit cannot monitor and control the transfer, and the requested data. The obtained data are then stored outside the infected computer. The network cards seem to be the ideal solution in this case, as they have the hardware DMA into memory, and can immediately send the collected data over a network to a remote machine for the data analysis.

Our design follows this idea. We implemented a new method that allows us to dump the memory content using NDIS driver, and the network card (NIC). It would also be possible to replace the NDIS driver used in the attack from protocol driver to intermediate or miniport driver. Then it will be possible to setup the NDIS driver not only by the control program installed on the local machine, but also through the network. The other advantage is that we can write, using NDIS driver, specific functions for getting the physical address of many critical OS components and structures. Using this information we can check the exact part of physical memory corresponding with protected or monitored system components (IAT / EAT / SSDT / IDT/ IRP tables, kernel functions (using integrity checks) or key data structures.

In this way, new possibilities for forensic purposes, as well as rootkit detection over the network can be enabled. Although the current implementation using protocol driver is still far from optimal, the imposed restrictions (such as the necessity to load the modified driver itself into a system) limits the potential misuse of this technique for illegal activities. Unfortunately, the installation of the modified driver can change the contents of the memory, and could not be used if a sophisticated rootkit (which knows about these tools) is already installed. To overcome this limitation, we can e.g. load the driver when the operation system starts. The best solution would be to implement this kind of driver already as a part of the operating system, as is suggested in [10]. It would still be necessary (and very difficult in practice) to ensure that the modified NDIS driver is not misused by unauthorized entities, and to implement proper security mechanism to protect the driver (we plan to test SMM mode described in [14] or some integrity check solution). Our DMA-enabling driver can be implemented in practice as a part of the data collection agent used in a more complex security solution.

Taking into account all advantages of the solution (access to memory using DMA, sending contents of the memory to other computers, control of the driver through network), it seems to be a promising approach for enabling memory access. The ability can be really helpful for forensics analysis, malware and rootkit detection, or as a generic basis for data collection agents but it must be more deeply tested. We plan to continue the development of the driver, as well as to test its forensic applications, in the future.

5 Acknowledgment

The publication was supported by Grant VEGA 1/0173/13.

References

[1] Chris Riesh. Inside Windows Rootkits, 2006, online: www.thehackademy .net/madchat/vxdevl/library/Inside Windows Rootkits.pdf

[2] Sparks, Sherriand Butler, Jamie. Raising The Bar For Windows Rootkit Detection, PhrackMagazine Volume 0x0b, Issue 0x3d, 2005.

[3] Michael Davis. Hacking Exposed Malware & Rootkits, McGraw-Hill, United States, Copyright, 2009 , ISBN 0071591192 / 97800715 91195.

[4] Jesse D. Kornblum , Exploiting the Rootkit Paradox with Windows Memory Analysis, International Journal of Digital Evidence Fall 2006, Volume 5, Issue 1, online: www.ijde.org .

[5] Rutkowska, Joanna. Detecting Windows Server Compromises with Patchfinder 2. January, 2004 online www.invisiblethings.org/papers/ rootkits_detection_with _ pat-chfinder2.pdf

[6] Szor, P. The Art of Computer Virus Research and Defense. Addison-Wesley Professional, 2005, ISBN 0321304543.

[7] Carvey, H.. 2009. Windows Forensic Analysis DVD Toolkit. 2. Edition. Syngress. June 11, 2009. ISBN-13: 978-1597494229, ASIN: 1597494224.

[8] Boileau, A.. 2006. "Hit By A Bus: Physical Access Attacks with Firewire" Security - As-sessment.com, Ruxcon, 2006. [cit. 2011–05–25]. Online: http://www.storm.net.nz/static/files/ab_firewire_ rux2k6-final.pdf.

[9] Carrier, B., Grand J.: A Hardware - Based Memory Acquisition Procedure for Digital In-vestigations. In Digital Investigation Journal. February 2004.

[10] N. L. Petroni, Jr., T. Fraser, J. Molina, and W. A. Arbaugh, "Copilot-a coprocessor-based kernel runtime integrity monitor," in SSYM'04: Proceedings of the 13th conference on USENIX Security Symposium. Berkeley, CA, USA: USENIX Association, 2004, pp. 13–13.

[11] A. Baliga, V. Ganapathy, and L. Iftode, "Automatic inference and enforcement of kernel data structure invariants," in ACSAC '08: Proceedings of the 2008 Annual Computer Security Applications Conference. Washington, DC, USA: IEEE Computer Society, 2008, pp. 77–86.

[12] DUFLOT, Loïc, Yves-Alexis PEREZ, Guillaume VALADON a Olivier LEVILLAIN. Can you still trust your network card?. In: Agence nationale

de la sécurité des systèmes d'information, 2010 [cit. 2012–05–16]. Online: http://www.ssi.gouv.fr/IMG/pdf/csw-trustnetworkcard.pdf

[13] DELUGRÉ, Guillaume. Closer to metal: Reverse engineering the Broadcom NetExtreme's firmware. In: Sogeti ESEC Lab [online]. 2010 [cit. 2012–05–16]. Dostupné z: http://esec-lab.sogeti.com/dotclear/public/publications/10-hack.lu-nicreverse_slides.pdf

[14] J. Wang, F. Zhang, K. Sun, A.Stavrou. Firmware-assisted Memory Acquisition and Analysis tools for Digital Forensics, (SADFE), 2011 IEEE Sixth International Workshop, 2011

[15] PCA USA. NDIS developer's reference, 2012 [cit. 2012–05–16]. Online: http://ndis.com

[16] ManTech Memory DD (MDD) released under GPL by Mantech International http://sourceforge.net/projects/mdd/

Biography

Stefan Balogh has been an assistant professor at the Slovak University of Technology Faculty of Electrical Engineering and Information Technology, since 2007. He teaches classes in Information security, Communication protocols and Computer crime. He is completing his Ph.D. in information studies, where his research interest are in the areas of the forensic memory analysis, Cryptology and Behavior-Based Malware Detection.

Making Static Code Analysis More Efficient

Pomorova O.V. and Ivanchyshyn D.O.

System Programming Department, Khmelnytskyi National University, Instytutska Str. 11, Khmelnytskyi, 29016, Ukraine, E-mail: o.pomorova@gmail.com, dmytro_ivanchyshyn@ukr.net

Received 19 March 2014; Accepted 27 April 2014;
Publication 2 June 2014

Abstract

Modern software is a complex high-tech product. Users and customers put forward a number of requirements to such products. Requirements depend on software purpose. However, reliability, fault tolerance, security and safety requirements are topical for all software types. One of the approaches for realization of such requirements in the implementation stage of software life cycle is a static source code analysis (SCA). The efficiency assessment task of the SCA tools is an actual problem. This paper presents the method of the efficiency evaluating of the software static source code analysis. It allows increasing the quality and reliability of software in general. The result of this work is a method of efficiency improving at the debugging stage and approach for selection of the static code analysis tools for software of various types.

Keywords: Source code analysis, security, vulnerabilities, weaknesses, static analysis efficiency, efficiency metrics.

1 Introduction

The company Veracode provides a "cloud" service for the analysis of software vulnerabilities. Every year it presents a detailed analysis of the vulnerabilities, which were discovered in software. Veracode State of Software Security (SoSS) Report Volume 5 examines data collected over an 18 month period from January 2011 through June 2012 from 22,430 applications [1]. This report

Journal of Cyber Security, Vol. 3 No. 1 , 77–88.
doi: 10.13052/jcsm2245-1439.315

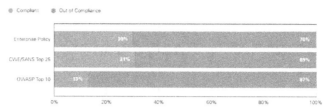

Figure 1   Compliance with Security Policies upon First Submission

examines application security quality, remediation, and policy compliance statistics and trends. It shows that 70% of applications failed to comply with enterprise security policies on first submission. Web applications are assessed against the Open Web Application Security Project (OWASP) Top 10 and only 13% complied on first submission. Non-web applications are assessed against the Common Weakness Enumeration (CWE/SANS) Top 25 and 31% complied on first submission. Only 30% of applications complied with enterprise defined policies (Figure 1). Therefore, the vast majority of applications are returned to the testing stage for further debugging.

Detection and correction of software defects (debugging) are two of the most difficult and time-consuming stages in the process of software development. Up to the 95% of debugging time is spent on the detection of defects and only 5 % is spent for defect correction [2]. So one of the actual issues for today is improving of the efficiency of software error detection. Reducing the time for identification of defects in software source code will significantly decrease the general time and resources that are spent in the debugging stage of the software lifecycle.

One advantage of our approach is the ability to consider and to identify actual vulnerability to a particular type of software at the implementation stage. It allows us to increase the quality and reliability of software in general. This paper presents an analysis of the weaknesses classification and identification problems and describes the ability to apply such information in the realization stage of the software lifecycle. The result of this work is an efficient method of improving the debugging stage and an approach for the selection of static code analysis tools for software of various types.

2 Static Code Analysis (SCA)

One of the methods of source code verification is static code analysis. SCA is the process of evaluating a system or component based on its form, structure, content, or documentation [3]. From a software assurance perspective, static analysis addresses weaknesses in program code that might lead to vulnerabilities. SCA is performed without actually executing programs and can be applied on the early stages of software lifecycle. Such analysis may be manual, as in code inspections or automated through the use one or more tools. Automated static code analyzers typically check source code but there is a smaller set of source code analyzers that check byte code and binary code. There are especially useful when source code in not available. Static code analyzers are used to uncover hard to find implementation errors before run-time, since they may be even more difficult or impossible to find and assess during execution. These tools can discover many logical, safety and security errors in an application without the need to execute the application.

2.1 Static Security Analysis (SSA)

One of the SCA categories is static security analysis (SSA). It attempts to identify errors and security weaknesses through deep analysis of source code [4]. SSA is primarily aimed at developers and QA engineers who wish detect software defects early in the development cycle in order to reduce time and cost. It also assists developers in hardening their application against security attack. SSA provides an effective way to discover defects, especially in code that is hard to exercise thoroughly with tests.

Many coding errors and patterns of unsafe usage can be discovered through static analysis. The main advantage of static analysis over dynamic analysis is that it examines all possible execution paths and variable values, not just those that are provoked during testing. This aspect of static analysis is especially valuable in security assurance, since security attacks often exercise an application in unforeseen and untested ways. SSA can detect different error conditions: buffer overflows and boundary violations, misuse of pointers and heap storage, memory leaks, use of uninitialized variables and objects, unsafe/incorrect use of functions, etc.

2.2 Static Code Analysis Efficiency

Static code analysis efficiency is a complex property that reflects the quality of the results, the degree of automation of the analysis and the complexity of its organization, resource-intensive, applicability to different class and size of programs [5]. The obtained results (based on the number of true positive (TP), false positive (FP), and false negative (FN) in the analyzer's reports) (2.1, 2.2). Other parameters are the importance of revealed defects, the properties of programming languages, type of software, features of the software algorithm and coding style, the impact of environment and external influences.

The F_score metric provides weighted guidance in identifying the most efficient static analysis tool by capturing how many of the weaknesses were found (true positives) and how much noise (false positives) was produced. An F_score is harmonic mean of the Precision and Recall values (2.3):

$$Precision = \frac{TP}{TP + FP} \tag{1}$$

$$Recall = \frac{TP}{TP + FN} \tag{2}$$

$$F\_Score = 2 \times \frac{Precision \times Recall}{Precision + Recall} \tag{3}$$

These parameters are universal and can be applied both for SCA and for SSA efficiency assessment.

2.3 Static Security Analysis Tools

SSA has a number of usage features. The security Department of the U.S. National Institute of Standards and Technology provides a list of SCA tools that can be used for detecting and reporting weaknesses that can lead to security vulnerabilities. The following SCA tools were select for this paper: Gimpel PC Lint ($389), PVS – Studio ($4585), Red Lizard Goanna Studio ($999), and CppCheck (freeware).

Three sets of test samples for quality of the SCA assessment were selected. The first set was taken from the website of the U.S. department of Homeland Security. The second set was taken from the website of the Security Department of U.S. National Institute of Standards and Technology. They included test samples designed specifically for SCA tools testing [6]. The examples are small, simple C/C++ programs, each of which is meant to evaluate some specific aspect of a security scanner's performance. The third test set contained applications with only one type of defect. The widespread defect in the source code CWE (Common Weakness Enumeration) 476 CWE null pointer dereference was selected. Each of the tools has a rule for identifying of such class of defects. The third test set is aimed at detection the differences between the methods used by developers to identify the stated defects. Table 1 shows the F _score metric results obtained for the three test sets.

Table 1 SCA Results

F_score	CppCheck	PVS-Studio	Goanna	PC-Lint
Test set 1	0,74	0,39	0,21	0,44
Test set 2	0,31	0,31	0,35	0,33
Test set 3	0,75	0,67	0,67	0,67

The efficiency of SCA tools varies significantly and changes greatly for different test sets. In addition, the list of defects detected by each analyzer was different. Modern investigations of other SCA tools efficiency also shows similar results [7, 8]. Therefore:

The efficiency of the tools depends on the usage scenario
SCA tools don't identify all defects in existing software

Consequently, it is hard for software developers and QA managers to choose one of the available SCA tools. In addition, tools have to be up to date with respect to the spectrum of threats, weaknesses, vulnerabilities and long-term costs.

2.4 Improving of the SCA Efficiency

Today there are a number of investigations on improving the efficiency of SCA. There are a number of regulations to ensure the quality of the static analysis of the application in the development of military software. Michael Howard proposes a list of recommendations for improving SCA efficiency [9]:

Multiple tools should be used to offset tool biases and minimize false positives and false negatives and minimize false positives and false negatives.
Analysts should pay attention to every warning and error.
Warnings from multiple tools may indicate that the code needs closer scrutiny (e.g. manual analysis).
Code should be evaluated early, preferable with each build, and re-evaluated at every milestone.

In addition, analysts should make sure that code reviews cover the most common vulnerabilities and weaknesses, such as integer arithmetic issues, buffer overruns, SQL injection, and cross-site scripting (XSS). Sources for such common vulnerabilities and weaknesses include the Common Vulnerabilities and Exposures (CVE) and Common Weaknesses Enumeration (CWE) databases, maintained by the MITRE Corporation. MITRE, in cooperation with the SANS Institute, also maintains a list of the "Top25 Most Dangerous Programming Errors" that can lead to serious vulnerabilities. Static code analysis tool and manual techniques should at a minimum address this Top 25. The better static code analysis tools are expensive. Other ways to improving the efficiency of SCA is to use multiple tools to offset tool biases and minimize false positives and false negatives. However, this is cost prohibitive.

The practical implementation of the recommended steps has a number of problems. Different SCA tools has different defects classification systems. It is hard to compare results obtained from different tools. Also, the bases of some analyzers do not cover the most common vulnerabilities and weaknesses. All of which make it impossible to fully use all benefits of the static analysis technology.

3 An Efficient Method for Static Code Analysis

3.1 Problems with Weaknesses Classifications

Every few years a «Top 25 Most Dangerous Software Errors» is constructed from the CWE. This is a list of the most widespread and critical errors that can lead to serious vulnerabilities in software. The higher a defect is located in the CWE ranking, the more important its detection by SCA tools. However, the rating is developed for all existing types of software, which complicates its practical application. For example, currently the most important is weaknesses in the list is CWE-89 (cross-site scripting). This problem is related to web services and it is not relevant to application or system software. Thus, for a particular type of software it is appropriate to apply special an importance rating of the defects (Table 2).

One approach to solving this problem is to carry out an analysis of NVD (national vulnerabilities database). The task was to identify the most common and critical defects for certain categories of software. NVD is the U.S. government repository of standards based vulnerability management data represented using the Security Content Automation Protocol (SCAP). NVD includes databases of security checklists, security related software flaws, misconfigurations, product names, and impact metrics. Figure 2 shows the number and criticality of all weakness obtained from NVD database. Table 2 presents weaknesses ratings with different criteria: relevance, most important weaknesses for all types of software, relevance only for operating systems.

Table 2 The Most Important Weaknesses

No	Top CWE 2011	Relevance	Most important	Relevance in OS
1	CWE-89	CWE-79	CWE-119	CWE-399
2	CWE-78	CWE-119	CWE-89	CWE-20
3	CWE-120	CWE-89	CWE-264	CWE-119
4	CWE-79	CWE-264	CWE-79	CWE-264
5	CWE-306	CWE-20	CWE-20	CWE-189
6	CWE-862	CWE-399	CWE-94	CWE-200
7	CWE-798	CWE-94	CWE-399	CWE-362
8	CWE-311	CWE-22	CWE-22	CWE-94
9	CWE-434	CWE-200	CWE-189	CWE-16
10	CWE-807	CWE-189	CWE-200	CWE-310
11	CWE-250	CWE-287	CWE-287	CWE-287
12	CWE-352	CWE-352	CWE-352	CWE-79
13	CWE-22	CWE-310	CWE-310	CWE-255
14	CWE-494	CWE-255	CWE-255	CWE-22

Criticality in CWE — CWE type	1	2	3	4	5	6	7	8	9	10
CWE-79	0	90	118	3847	4	6	14	1	2	0
CWE-119	0	16	10	161	343	63	609	598	1457	777
CWE-89	0	1	0	4	25	48	421	3009	6	9
CWE-264	3	129	35	409	646	247	540	363	167	128
CWE-20	2	44	38	298	518	225	307	362	288	132
CWE-399	0	32	25	209	503	30	197	366	306	140
CWE-94	0	0	9	40	69	35	428	550	460	112
CWE-22	0	0	7	118	451	97	285	341	83	47
CWE-200	6	153	38	260	714	22	35	43	6	11
CWE-189	2	9	12	70	177	12	177	128	240	108
CWE-287	0	5	7	41	115	73	105	261	38	71
CWE-352	0	0	4	70	19	44	385	12	10	0
CWE-310	2	51	13	69	147	39	31	56	13	23
CWE-255	1	48	5	38	78	11	31	49	12	56
CWE-59	8	14	54	39	20	26	154	6	3	1
CWE-16	0	11	9	40	65	28	33	37	15	15
CWE-362	4	15	3	30	60	53	57	9	11	2
CWE-134	0	2	0	8	19	7	24	35	23	26
CWE-78	0	0	0	2	1	0	9	17	26	19

Figure 2 Number and Criticality of Weaknesses from NVD Database

For operating systems, the distribution of weaknesses varies depending on the operating system: Windows, RedHat, Novel, Solaris, Apple or others (Figure 3).

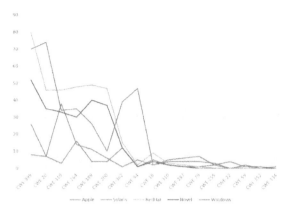

Figure 3   Distribution of Weaknesses for OSs

Figure 3 shows that for different types of software the distribution of defects varies significantly. It is reasonable to use different SCA tools for particular types of software.

3.2 Improving the Efficiency of SCA

The first step is to determine the type of software (ST) (Figure 4). The analysis of the NVD database shows that for different types of software there is a different list of most widespread weaknesses.

The second step, the informational sources (IST) are chosen. From the set of sources that contain information about the defects identified in the software, the most appropriate sources for subject area are selected. For example, for SSA such source can be the NVD.

Development of the defects list for a given type of software (STw) is the third step. Based on the analysis of the prevalence and criticality of defects given in IST the rating of the most important of defects is formed. The list includes weaknesses that have to be identified by SCA tools.

The fourth step is formation of the test samples set (TST) that cover the list of weaknesses (STw). For each weakness the test samples set that will be used for SCA checking is formed. For example, for buffer overflow weakness CWE-120 a set of test samples are given in Table 3.

The fifth step is the static analysis of test samples set by SCA tools. The result of the stage is a log-file with defects that have been detected.

Parsing of the log-files is the sixth step. The number of false positive, false negative, and true positive, is calculating in this stage.

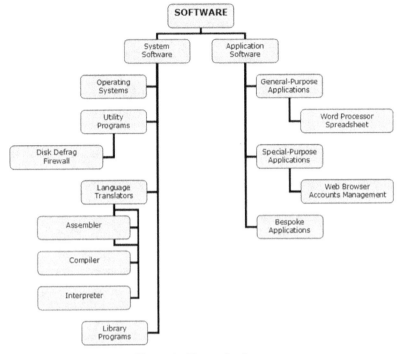

Figure 4   Types of software

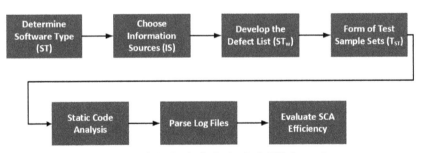

Figure 5   Method of the Static Code Analysis Efficiency Assessment

The last step is an analysis of the efficiency of the SCA. This step includes the calculation of Precision, Recall and F_score metrics. They are based on the number of true positives (TP), false positives (FP), and false negatives (FN) in the analyzer's report. The F_score metric is a final result that can be used for different SCA tools comparison. Figure 5 presents the steps for our method.

Table 3 Set of tests for buffer overflow weakness

Error type	Error name	CWE number
Arrays	Direct overflow	CWE-119
	Off-by-one errors	CWE-193
	Unbounded copy	CWE-120
Strings	Direct overflow	CWE-119
	Null termination	CWE-170
	Off-by-one errors	CWE-193
	Truncation error	CWE-222
	Unbounded copy	CWE-120
	Strcpy_check	
	strcat_check	
	gets_check	
	strncpy_check	
	strncat_check	
	fgets_check	
Integer	Overflow	CWE-190
	Sign errors	CWE-195
	Truncation errors	CWE-197

The user choses only the type of software and the information sources. The result is the most appropriate static code analysis tool for concrete software project.

4 Conclusion

Our investigation shows the existence of significant differences in the results produced by static code analysis tools. The efficiency of modern SCA tools depends on the usage scenario. In terms of software security, it depends on consideration of information about actual vulnerabilities. Such data can be obtained from databases like the NVD. Therefore, it is advisable to integrate information about known vulnerabilities and weaknesses in source code into the code analysis process.

Consequently, for software quality improvement and security assurance it is necessary to pay attention to the problem of the choosing appropriate SCA tools. The efficiency of a tool has to be evaluated and the information about actual vulnerabilities ought to be taken into account even at the software lifecycle stage of implementation. The proposed method provides a solution to this problem.

The problems of informational sources choosing for particular type of software, test sets development for defects checking and automatic parsing of log files are needed further investigations.

References

[1] Veracode Inc., State of Software Security Report: Volume 5, April 2013, 44 p.
[2] Ian Sommerville, Software Engineering (9th Edition), 2010.
[3] R. Lopes, D. Vicente, N. Silva. Static Analysis tools, a practical approach for safety-critical software verification. Critical Software SA Parque Industrial de Taveiro. Coimbra, Portugal, 2009, 12 p.
[4] Intel Corporation, Improve C++ Code Quality with Static Security Analysis (SSA), 2013, 11 p.
[5] National Security Agency Center for Assured Software. On Analyzing Static Analysis Tools. July, 2011.
[6] Build Security In. Source Code Analysis Tools - Example Programs: https://buildsecurityin.us-cert.gov/bsi/articles/tools/code/498-BSI.html
[7] Thomas Hofer. Evaluating Static Source Code Analysis Tools, School of Computer and Communications Science, Ecole Polytechnique Federal de Lausanne, March 12, 2010
[8] R. Plösch, A. Mayr, G. Pomberger, M. Saft. An Approach for a Method and a Tool Supporting the Evaluation of the Quality of Static Code Analysis Tools. Proceedings of SQMB 2009 Workshop, SE 2009 conference, Kaiserslautern, Germany, July 2009.
[9] Howard, M. A Process for Performing Security Code Reviews, IEEE Security & Privacy, July-August 2006, pp. 74–79.

Biographies

Oksana Pomorova. Doctor of Technical Science, Head of System Programming Department, Full Professor in Khmelnitsky National University (Ukraine). Received the PhD degree in Kyiv Institute of Automatics (2002), the degree Doctor of Technical Science in 2008 in the National University "Lviv Polytechnic" (Ukraine), specialty 05.13.13 - "Computers, Systems and Networks". IEEE member from 2005. *Teaching* - Computer Modeling, Technology of Software Design, Artificial Intelligence Systems. *Guest lectures*: Department of Computer Systems and Networks, Yuriy Fedkovych Chernivtsi National University (Ukraine); Kielce University of Technology (Poland). *Research Interests*: Intelligent Methods and Means of Computer Systems Diagnosing, Quality Assessment of Critical Software; Modeling and Design of Knowledge Bases for Testing and Diagnosing Specialized Computer Systems.

Dmytro Ivanchyshyn. He defended his master's thesis in Khmelnitsky National University. Now studying in postgraduate at the Faculty Programming, Computer and Telecommunication Systems and working as teacher trainee of System Programming Department. His research interests are: Software Quality Assurance and Testing, System Security

www.ingramcontent.com/pod-product-compliance
Lightning Source LLC
LaVergne TN
LVHW012333060326
832902LV00011B/1863